"十四五"职业教育国家规划教材

计算机专业英语

第六版

新世纪高职高专教材编审委员会 组编
主 编 卢川英 邵奎燕
副主编 张广军 迟晓曼

大连理工大学出版社

图书在版编目（CIP）数据

计算机专业英语 / 卢川英, 邵奎燕主编. -- 6版
. -- 大连 : 大连理工大学出版社, 2022.1（2025.1重印）
新世纪高职高专计算机应用技术专业系列规划教材
ISBN 978-7-5685-3554-0

Ⅰ. ①计… Ⅱ. ①卢… ②邵… Ⅲ. ①电子计算机 - 英语 - 高等职业教育 - 教材 Ⅳ. ①TP3

中国版本图书馆CIP数据核字(2022)第013170号

大连理工大学出版社出版

地址: 大连市软件园路 80 号　邮政编码: 116023

营销中心: 0411-84707410 84708842　邮购及零售: 0411-84706041

E-mail: dutp@dutp.cn　URL: https://www.dutp.cn

大连永盛印业有限公司印刷　大连理工大学出版社发行

幅面尺寸:185mm × 260mm　印张: 14.75　字数: 339千字

2003年9月第1版　2022年1月第6版

2025年1月第7次印刷

责任编辑: 李　红　责任校对: 马　双

封面设计: 张　莹

ISBN 978-7-5685-3554-0　定价: 49.80 元

前　言

《计算机专业英语》（第六版）是“十四五”职业教育国家规划教材、“十三五”职业教育国家规划教材、“十二五”职业教育国家规划教材，也是新世纪高职高专教材编审委员会组编的计算机应用技术专业系列规划教材之一。

众所周知，英语在计算机领域及IT行业有着举足轻重的作用。新世纪信息技术的专业人才，不仅需要掌握扎实的专业基础知识和基本技能，还应当具备一定的英语运用能力，高职专业院校的学生更应注重英语运用能力的培养。因此，我们从高职院校人才培养要求出发，编写了这本独具特色的《计算机专业英语》（第六版）教材。

本教材特点如下：

1.树宗旨——将立德树人课程思政融入教材编写全过程

本教材以习近平新时代中国特色社会主义思想为指导，全面贯彻党的教育方针，以培养担当民族复兴大任的时代新人为着眼点，立足培养学生的工匠精神、创新思维、操作技能、较强的就业能力和可持续发展能力。在教材“理论学习”和“实践学习”模块的内容选取上，大量选用我国自主研发产品作为素材，对学生进行爱国主义教育，厚植爱国主义情怀，引导学生肩负我科技强国、网络强国的使命；在“职场英语”模块中全程注入“工匠精神”理念，引导学生除了要具备专业技能之外，还要具备高尚的职业道德和职业精神，将立德树人课程思政融入教材编写全过程。

2.强专业——教材内容与该学科人才培养目标完全吻合

本教材遵循《国家职业教育改革实施方案》，聚焦提升技术技能人才培养质量，服务我国先进电子信息行业，将ICT领域新兴技术（如大数据、物联网、5G、人工智能等）纳入教材内容，让学生在全英环境下完成专业知识和操作技能的学习。教材中涵盖的计算机专业知识全面、准确，与该学科专业人才培养目标完全吻合，充分体现专业英语的核心要素——“专业”。

3.创特色——以实际项目为载体、以学生就业为导向、以能力培养为本位、以训练为手段

本教材从职业教育人才培养要求出发，按照“理论学习、实践练习、职场英语”模块化设计思路进行教材设计与编写，特色鲜明。

（1）以实际项目为载体的设计理念

让学生在全英语环境下完成从“计算机硬件软件安装—网络组建与连接—信息查询分析—搭建网上商城—创建商品数据库—网络安全防护—网上交易—信息追踪—人工智能”这个涵盖了ICT专业主要工作任务的总项目,项目设计环环相扣，独具匠心，突出专业性、实用性

和创新性。

（2）以学生就业为导向设计教学内容

本教材为实现理论与实践、知识与技能的有机整合，每个子项目采用模块化设计思路，包含三个模块，分别是Theoretical Learning（理论学习）模块、Practical Learning（专业任务）模块、Occupation English（职场英语训练）模块。学生通过Theoretical Learning（理论学习）模块，在老师的指导下，在实验室完成相应的Practical Learning（专业任务）模块以巩固和检验专业英语知识的学习情况；并针对行业职业岗位需求，在每个子项目中增加了Occupation English（职场英语）模块，模拟实际的工作情境进行未来工作岗位职场英语对话，使学生在工作岗位中（或求职过程中）能够真正做到会听、会说、会用。

（3）以能力培养为本位，以训练为手段

针对在科技英语阅读过程中普遍存在的“病症”，将每个单元分为几个训练模块，如句子主干训练、关键词训练、猜词训练等，每个训练模块训练目标明确，将英语语言学习中的阅读方法贯穿于专业英语的教学中，有针对性地加以训练，使学生能够切实地掌握科技英语的阅读技巧，快速提高科技英语的阅读能力。

（4）版式新颖，取材与时俱进

本教材内容取材多来源于英文网站和英文资料的原版文章，力求反映计算机方面的新知识和新技术。文中生词的注释不像以往放在文章结束的地方，而是标示在正文的一侧，以便读者阅读。每个单元的结尾还增加了构词法，帮助读者掌握英文中最常用的构词方法，使读者能够通过一词掌握多词，迅速扩充词汇量，克服英语学习中单词记忆困难的瓶颈。

4.融数字——齐备的数字资源促进课程信息化教学模式创新

本教材围绕深化教学改革和“互联网+职业教育”发展需求，注重满足分类施教、因材施教的需要，开发了全书音频等系列配套数字资源，将信息技术与教材完美融合，为学生泛在自学与教师课堂授课构建了完善的辅助平台，有利于信息化教学模式的创新。

本教材由吉林交通职业技术学院卢川英、邵奎燕任主编，吉林交通职业技术学院张广军、迟晓曼任副主编，吉林网格信息技术有限公司马贺参与编写。具体分工如下：项目1、3、7由卢川英编写；项目4、9、10由邵奎燕编写；项目5、6由张广军编写；项目2、8由迟晓曼编写；各项目的Occupation English模块由马贺编写，本教材由邵奎燕统稿。

另外，尽管我们在这本《计算机专业英语》教材体系设计方面做了创新和改革，希望对以后的高职《计算机专业英语》类教材的编写能起到抛砖引玉的作用，但由于作者水平有限，时间仓促，不足和疏漏在所难免，恳请各相关单位和读者在使用本教材的过程中给予关注，并能进行批评指正，并将意见及时反馈给我们，以便我们能及时修订和改进，促使我们不断地共同进步，在此编者深表感谢！

编　者

2022年1月

所有意见和建议请发往：dutpgz@163.com
欢迎访问职教数字化服务平台：https://www.dutp.cn/sve/
联系电话：0411-84707492　84706104

Contents

Navigation

我们在阅读科技英语的过程中，普遍存在若干“病症”，下面列举一些主要“病症”并辅以“对症良药”。

病症之一：科技英语书中的句子有别于我们在其他英语书中常见到的句子，其从句、插入语、倒装句相对较多，不少初学者很不习惯，以致常常读不懂句子。

对症良药：尝试快速分辨句子的主、谓、宾或表语，迅速读出句子的主干，掌握大意。

Skill One：主要训练分辨句子的主、谓、宾，把握句子主干，掌握句子以及段落的大意。主语用____表示，谓语用___表示，宾语用﹏﹏表示，从句引导词用□表示。

病症之二：普通英语注重“精读”，即“word by word”“sentence by sentence”。因而大多数读者在阅读科技英语时都是在逐字逐句地读，极大地影响了阅读速度和阅读效果。

对症良药：专业英语注重的是“泛读”，抓住文章或段落的关键词（keyword），通过关键词来了解句子乃至段落的意思，再根据实际情况围绕关键词进行精读或慢读。

Skill Two：主要训练快速抓住关键词来掌握句子及段落的大意，提高阅读速度。

病症之三：我们一般都有遇到生词就查词典的习惯，一篇文章总有若干生词，几次查下来，阅读速度大打折扣，且会极大地影响阅读兴趣。

对症良药：①培养猜词技能。遇到生词姑且放过，先不理会，继续往下读，然后根据前后文猜测出该词的意思。②学习词根构词法。英语中很多单词都是由词根衍变而来的，记住词根会大大提高背单词的效率。

Fast Reading：主要训练猜词技巧。每个单元后的Word Building将帮助你快速掌握构词法。

病症之四：我们学了很多年的英语，但在求职过程中或将来的工作岗位上，可能根本不会用。

对症良药：在英语学习的过程中注重多读、多听、多说。尤其是针对职业的岗位需求，模拟实际的工作情境进行未来工作岗位职场英语对话。

Occupation English：主要训练计算机专业学生未来工作岗位职场英语对话。

功夫不负有心人，相信读者掌握了正确的科技英语的学习方法后，再经过一段时间的训练，专业英语的水平会有极大的提高。

Project One

Assembling Computers

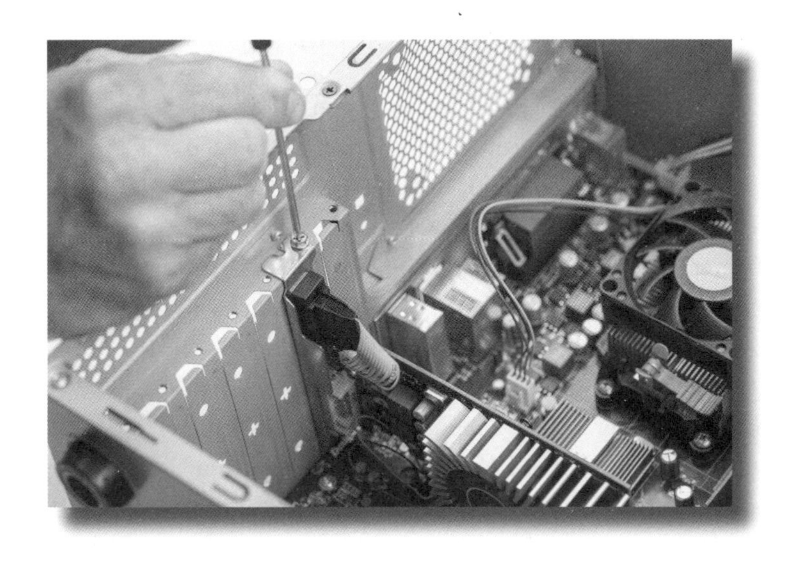

Part A Theoretical Learning

Part B Practical Learning

Part C Occupation English

Part A Theoretical Learning

Training Target

In this part, our target is to improve the speed of reading professional articles and the comprehension ability of the reader. We have marked specialized key words and some flexible sentences. Try to grasp the main idea of each paragraph.

Skill One A Short Introduction to Computers

A computer is just a machine, but a computer system consists of two main elements: machine and **program**. Like a person, a computer system is composed of two parts: the bone — **hardware** and the soul — **software**. The central idea of a computer system is that input is processed into output. Input is the data which is entered into the computer, and output is the result of processing done by the computer, usually printed out or displayed on the screen.

program ['prəʊgræm] *n.* 程序
hardware ['hɑ:dweə(r)] *n.* 硬件
software ['sɒftweə(r)] *n.* 软件

Let's get closer to the computer from the basic **components**. When we're talking about computers, such images as Pic 1.1 will appear in our mind: a **display screen** known as the basic **output device**, a **keyboard** usually together with a **mouse** as the basic **input device**, and a **cabinet** known as a machine box.

component [kəm'pəʊnənt] *n.* 部件
display screen 显示器
output device 输出设备
keyboard ['ki:bɔ:d] *n.* 键盘
mouse [maʊs] *n.* 鼠标
input device 输入设备
cabinet ['kæbɪnət] *n.* 匣子，机箱

display screen
machine box
mouse
keyboard

Pic 1.1 A Computer's Basic Components

Skill One

With the development of science and technology, the modern computer becomes more and more **flexible**, and the hardware family becomes stronger and stronger. A lot of new **peripherals** have appeared. These peripherals can be classified into two groups — input devices and output devices.

flexible ['fleksəbl]
adj. 灵活的
peripheral [pə'rɪfərəl]
n. 外围设备

Input devices (Pic 1.2), as the name suggests, are any hardware components that allow you to put the data, programs and commands into the computer. One of the most important input devices is the keyboard. Users can type in texts or enter keyboard commands using the keyboard. Another device that can be used to input data is **scanner**. This electronic device is used to transfer an image such as text, or pictures into the computer. The most useful **pointing device** is a mouse, which allows the user to point to elements on the screen. There are some other input devices, such as **microphone, PC camera, digital camera, joystick, graphics tablet** and **light pen.**

scanner ['skænə(r)]
n. 扫描仪
pointing device 定点设备
microphone ['maɪkrəfəʊn]
n. 麦克风
PC camera 电脑摄像头
digital camera 数码相机
joystick ['dʒɔɪstɪk]
n. 控制杆
graphics tablet 图形输入板
light pen 光笔

Pic 1.2 Input devices

Output devices (Pic 1.3) are devices that let you see what the computer has accomplished. Several devices are used to display the output from a computer. **The favorite monitor is the LCD, which is slim and takes up little space, and displays texts and images with greater clarity.** Another important output device is the printer,

LCD (Liquid Crystal Display)
abbr. 液晶显示器

which allows the user to copy the data in the computer onto the paper. **Speakers** and **earphones** allow the listener to hear **audio** data through the computer, such as speech or music. There are some other output devices, such as **projector** and **facsimile machine**.

speaker ['spi:kə(r)] *n.* 扬声器
earphone ['ɪəfəʊn] *n.* 耳机
audio ['ɔ:diəʊ] *n.* 音频
projector [prə'dʒektə(r)] *n.* 投影仪
facsimile machine 传真机

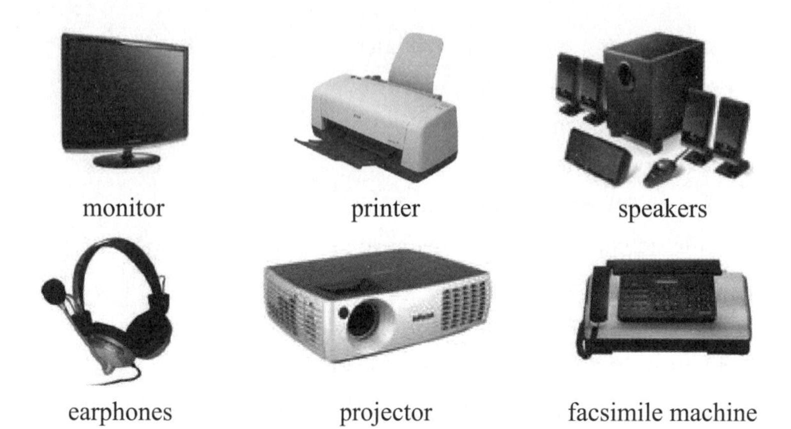

Pic 1.3 Output Devices

All the components of a computer that we can see and feel are hardware. They work together to help us with our daily work.

Do you know how a computer can manage so many devices? The real secret lies in the machine box. When we take the cover off a small computer and look inside, the real computer appears in front of us and we will see a few circuit boards, some **wires** and some **cables**. In fact, the **motherboard** is the most important part in the machine box. Two main components on the motherboard are the **CPU** and the **memory**.

wire ['waɪə(r)] *n.* 导线
cable ['keɪbl] *n.* 电缆
motherboard ['mʌðəbɔ:d] *n.* 母板，主板

The CPU is sometimes referred to as the processor. It is the electronic device that interprets and carries out the basic instructions that operate the computer. The CPU is the control and data processing center of the whole computer system. You can simply regard it as a skillful **cook**. The only difference is just that the cook processes meat and vegetables, and makes them become delicious dishes. Here meat and vegetables are the input for the cook, and some dishes certainly are the output from the cook. Now turn back to our CPU: it can process the **digital data** from any input devices, and output them to an output device.

CPU (Central Processing Unit) *abbr.* 中央处理器
memory ['meməri] *n.* 内存
cook [kʊk] *n.* 厨师
digital data 数字数据

Just like the excellent cook we mentioned before, he must need a number of empty plates around him, which stored meat and vegetables to be processed by the cook, which is a great help for his cooking. And after dinner, the plates should be cleaned up. Memory stores information processed by the CPU. The **data stream** can flow from

data stream 数据流

the CPU into memory or on the contrary. Memory consists of **RAM** and **ROM**. Any information in RAM will be lost when the computer is turned off, just like the plates that are cleaned up by the cook.

Most of the devices connected to the computer communicate with CPU in order to carry out a task. The CPU controls the data flow on the **inner Bus**. There are three kinds of Buses used in our computer: **AB, DB** and **CB**. The most popular Bus to be used on a motherboard is a **PCI** Bus which is a peripheral component interface Bus.

The CPU uses storage to hold data, instructions and information for future use. Storage is also called secondary storage or **auxiliary storage**. Think of storage as a little cabinet used to hold **file folders**, and memory as the top of your desk. When you need a file, you can get it from the **filing cabinet** (storage) and place it on your desk (memory). When you finish a file, you return it to the filing cabinet (storage). The items in storage are **retained** even when power is off from the computer.

.End.

contrary ['kɒntrəri] *n.* 相反
RAM (Random Access Memory) *abbr.* 随机存取存储器
ROM (Read Only Memory) *abbr.* 只读存储器
inner Bus 内部总线
AB (Address Bus) *abbr.* 地址总线
DB (Data Bus) *abbr.* 数据总线
CB (Control Bus) *abbr.* 控制总线
PCI (Peripheral Component Interconnect) *abbr.* 外设部件互连
auxiliary storage 辅助存储器
file folder 文件夹
filing cabinet 文件柜
retain [rɪ'teɪn] *v.* 保留

Key Words

program *n.* 程序	hardware *n.* 硬件
software *n.* 软件	component *n.* 部件
display screen 显示器	output device 输出设备
keyboard *n.* 键盘	mouse *n.* 鼠标
input device 输入设备	cabinet *n.* 匣子，机箱
flexible *adj.* 灵活的	peripheral *n.* 外围设备
scanner *n.* 扫描仪	pointing device 定点设备
microphone *n.* 麦克风	PC camera 电脑摄像头
digital camera 数码相机	joystick *n.* 控制杆
graphics tablet 图形输入板	light pen 光笔
speaker *n.* 扬声器	earphone *n.* 耳机
audio *n.* 音频	projcetor *n.* 投影仪
facsimile machine 传真机	wire *n.* 导线
cable *n.* 电缆	motherboard *n.* 母板，主板
CPU *abbr.* 中央处理器	memory *n.* 内存
cook *n.* 厨师	digital data 数字数据

Key Words

data stream 数据流
contrary *n.* 相反
RAM *abbr.* 随机存取存储器
ROM *abbr.* 只读存储器
inner Bus 内部总线
AB *abbr.* 地址总线
DB *abbr.* 数据总线
CB *abbr.* 控制总线
PCI *abbr.* 外设部件互连
auxiliary storage 辅助存储器
file folder 文件夹
file cabinet 文件柜
retain *v.* 保留

参考译文 技能1 计算机简介

一台计算机只是一部机器，而一个计算机系统则包括两个要素：机器和程序。像人一样，计算机系统也由两部分组成：硬件（像人的骨架）和软件（像人的灵魂）。计算机系统的核心思想是把输入处理成为输出。输入是进入计算机的数据，而输出则是计算机处理的结果，一般由打印机打印或显示器显示出来。

让我们就计算机的基本构成来进一步认识计算机。当我们谈论计算机时，脑海中会出现如图1.1所示的画面：显示器是基本的输出设备，键盘和鼠标一起作为基本的输入设备，还有一个称为机箱的盒子。

随着科学技术的发展，现代计算机也越来越便捷，硬件家族也越来越强大，出现了许多新式外围设备。这些外围设备可以分为两类——输入设备和输出设备。

顾名思义，输入设备(图1.2)是让用户向计算机输入数据、程序和命令的硬件部件。最重要的输入设备之一是键盘。用户可以用它录入文本或输入键盘命令。可用来输入数据的另一种设备是扫描仪。这种电子设备可将文本、图片等影像传入计算机。最有用的定点设备是鼠标，它使用户能够指向屏幕上的内容。还有一些其他的输入设备，如麦克风、电脑摄像头、数码相机、控制杆、图形输入板和光笔。

输出设备（图1.3）可以让你看到电脑已经完成的内容。有几种设备可以用于计算机的输出显示。最受欢迎的显示器是液晶显示器，它体积小、占地少，能更清晰地显示文本和图像。另一种重要的输出设备是打印机，用户能用它将计算机中的数据拷贝到纸上。扬声器和耳机能使用户通过计算机听到演讲或音乐等音频数据。还有一些其他的输出设备，如投影仪和传真机。

计算机中所有我们看得见、摸得着的部件都叫作硬件。它们相互协作，帮助我们处理日常工作。

你知道计算机怎样管理这么多的设备吗？真正的秘密就在机箱里。当我们把小型的计算机的机箱盖取下并观察其内部的时候，一台真正的计算机就展现在我们面前了。我们会看到几块电路板、一些导线和一些电缆。事实上，机箱里最重要的部件是主板，主板上有两个主要的部件：中央处理器和内存。

中央处理器有时也被称为处理器，该电子设备用来解释并执行计算机的基本操作指令。中央处理器是整个计算机的系统控制和数据处理中心。你可以简单地将其当作一个熟练的厨师，不同之处只不过是厨师加工肉类和蔬菜，把它们变成美味的佳肴。这里肉类和蔬菜是输入给厨师的，而这些佳肴当然是从厨师那里输出的。现在回到我们的中央处理器，它可以处理从输入设备输入的数字数据，并将其输出到某种输出设备上。

正如我们刚才提到的那个优秀的厨师，在他周围一定需要许多空盘子，这些盘子用于存储他将要处理的肉类和蔬菜，这对他做菜有很大的帮助，而且在宴会之后，盘子就会被清洗干净。内存存储中央处理器所处理的信息。数据流可以从中央处理器流入内存或者从内存流入中央处理器。内存由随机存取存储器和只读存储器组成。当计算机关机时，在随机存取存储器中的信息将会丢失，正如同盘子被厨师清洗干净一样。

大多数连接到计算机上的设备通过与中央处理器通信来完成任务。中央处理器控制着内部总线上的数据流。计算机有三种总线：地址总线、数据总线和控制总线。在主板上最常用的总线是PCI总线，即外设部件互连总线。

中央处理器采用存储器保存将来要使用的数据、指令和信息。存储器也叫第二存储器或辅助存储器。你可以把存储器看作是一个用来存放文件夹的小柜子，把内存当作办公桌面。当你需要一份文件时，你可以从文件柜（存储器）中获取并把它放置在办公桌面（内存）上；当你用完这份文件后，你又可以把它放回文件柜（存储器）。存储器里的内容即使是在计算机断电时也能被保存。

Skill Two How Does a Computer Work?

When we talk about the computer, we usually meet the topic: motherboard (Pic 1.4). The main circuit board in a computer is called the motherboard. It is a flat board that holds all of the key elements that make up the "brain" of the system, including the **microprocessor** or CPU, RAM or primary memory, and expansion slots which are **sockets** where other circuit boards called expansion boards may be **plugged in**.

microprocessor [ˌmaɪkrəʊ'prəʊsesə(r)] *n.* 微处理器
socket ['sɒkɪt] *n.* 插座
plug in 插入

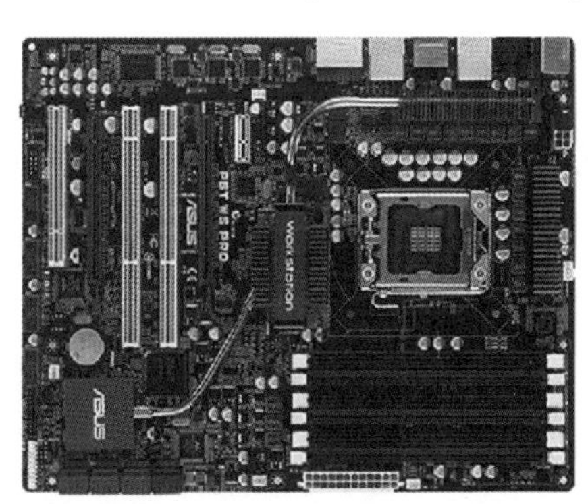

Pic 1.4 Motherboard

Skill Two

Let's use the system presented in Pic 1.5 to show you how a typical computer works. A computer is controlled by a stored program. So if we want to use a computer, the first step is to copy the program from diskette into memory. Now the **processor** can begin **executing** instructions; the data input from the keyboard is stored in memory. The processor processes the data and then stores the results back into memory. At last, we can get the results.

processor ['prəusesə(r)]
n. 处理器
execute ['eksɪkju:t]
v. 执行

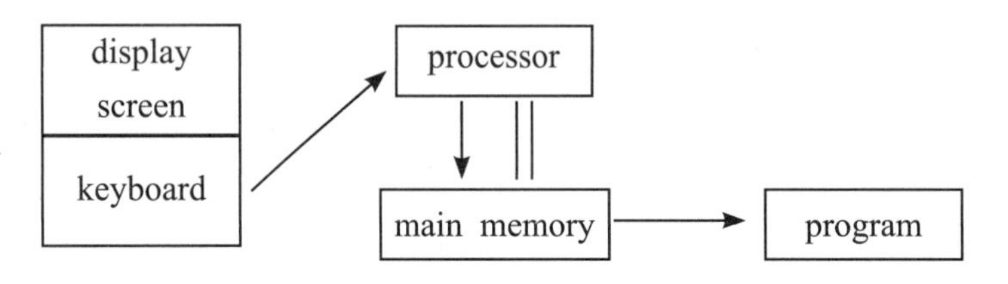

Pic 1.5 Computer System

Now we can see that a computer system consists of four basic components. An input device provides data. The data is stored in memory, which also holds a program. Under the control of the program, the computer's processor processes the data. The results flow from the computer to an output device. Let's introduce the system components one by one, beginning with the processor.

The processor, usually called the Central Processing Unit (CPU, Pic 1.6) or main processor, is the heart of a computer. It is the CPU that in fact processes or **manipulates** data and controls the rest part of the computer. How can it manage its job? The secret is software. Without a program to provide control, a CPU can do nothing. How can a program guide the CPU through the processes? Let's consider from the basic element of a program — instruction. An instruction is composed of two parts: an **operation code** and one or more **operands** (Pic 1.7). The operation code tells the CPU what to do and the operands tell the CPU where to find the data to be manipulated.

manipulate [mə'nɪpjuleɪt]
vt. 处理，操作

operation code 操作码
operand ['ɒpərænd]
n. 操作数

Pic 1.6 CPU

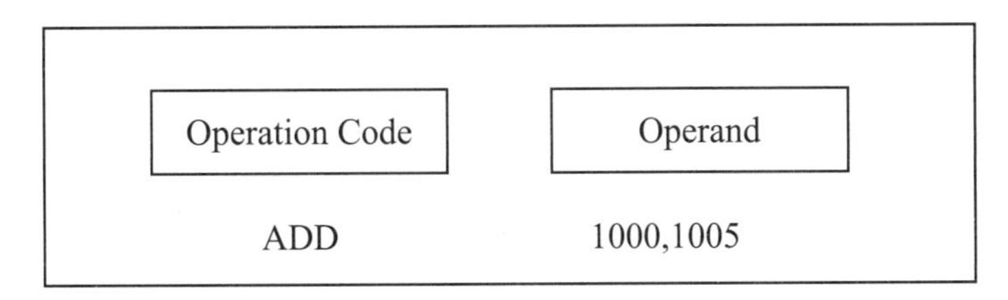

Pic 1.7 Instruction

The processor contains four major components (Pic 1.8): a clock, an instruction control unit, an arithmetic and logic unit (usually shortened to ALU) and several **registers**. The clock generates **precisely** timed **pulses** of **current** that **synchronizes** the processor's other components. Then the instruction control unit determines the location of the next instruction to be executed and fetches it from the main memory. The arithmetic and logic unit performs arithmetic operations (such as addition and subtraction) and logic operations (such as testing a value to see if it is true), while the registers are **temporary** storage devices that hold control information, key data and some **intermediate** results. Since the registers are located in the CPU, the processing speed is faster than the main memory. Then which is the key component to determine a computer's speed? It is the clock! In more detail, it is the clock's **frequency** that decides a computer's processing speed. When we buy a computer, we usually consider the main frequency first, and that means a clock's frequency.

register ['redʒɪstə(r)] *n.* 寄存器
precisely [prɪˌsaɪsli] *adv.* 精确地
pulse [pʌls] *n.* 脉冲
current ['kʌrənt] *n.* 电流
synchronize ['sɪŋkrənaɪz] *v.* 同步
temporary ['temprəri] *adj.* 暂时的
intermediate [ˌɪntə'mi:diət] *adj.* 中间的
frequency ['fri:kwənsi] *n.* 频率

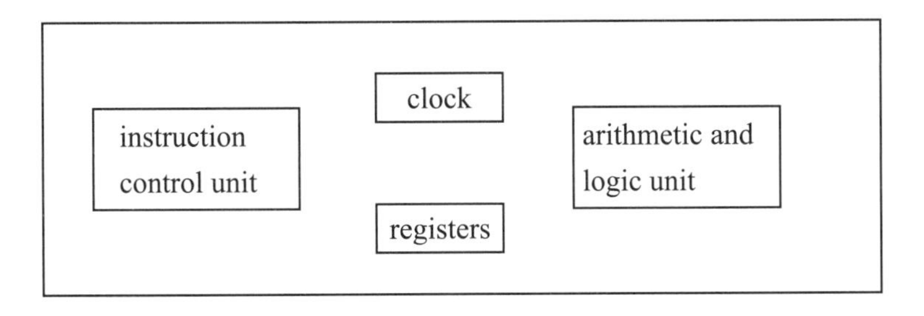

Pic 1.8 Processor's Four Major Components

Now we will talk more in detail about the Microprocessors and Central Processing Units. Microprocessors are central processing units etched on a tiny chip of silicon and, thus, are called microchips. Microprocessors contain many electronic **switches**, called transistors, which determine whether electric current is allowed to pass through. Transistors are the basic building blocks of microprocessors. A single microchip may contain millions of transistors. When electric current is allowed to pass through, the switch is on. This **represents** 1 bit.

switch [swɪtʃ] *n.* 开关
represent [ˌreprɪ'zent] *v.* 代表

If the current is not allowed to pass through, the switch is off. This represents a 0 bit. Different **combinations** of transistors represent different combinations of bits, which are used to represent special characters, letters, and digits.

combination [ˌkɒmbɪ'neɪʃn] *n.* 组合

.End.

Key Words

microprocessor *n.* 微处理器	socket *n.* 插座
plug in 插入	processor *n.* 处理器
execute *v.* 执行	manipulate *vt.* 处理，操作
operation code 操作码	operand *n.* 操作数
register *n.* 寄存器	precisely *adv.* 精确地
pulse *n.* 脉冲	current *n.* 电流
synchronize *v.* 同步	temporary *adj.* 暂时的
intermediate *adj.* 中间的	frequency *n.* 频率
combination *n.* 组合	

参考译文 技能2 计算机是如何工作的?

当我们谈及计算机的时候，我们通常会遇到这样一个话题：主板（图1.4）。计算机内的主要电路板叫作主板。它是一个平板，存放着所有组成系统“大脑”的关键元素，包括微处理器（或CPU）、RAM（或主存），还有一些扩展槽，它们是一些插口，可以将其他电路板（扩展板）插到里面。

让我们用图1.5所示的系统来说明一个典型的计算机是如何工作的。计算机是受存储程序控制的，所以，如果我们想使用计算机，第一步是将程序从磁盘上拷贝到主存中。现在，处理器可以开始执行指令了，从键盘输入的数据也存到主存中了。用处理器加工数据，然后把结果存回主存。最后，我们就得到了结果。

现在我们可以看到计算机系统由四个基本的部件组成：提供数据的输入设备、存放数据的存储器（程序也保存在内）、在程序控制下处理数据的处理器、输出结果的输出设备。让我们逐个介绍这些部件，先从处理器开始。

处理器通常被称为中央处理器（CPU）（图1.6）或主处理器，是计算机的心脏。事实上，是由CPU来处理或操作数据，并且控制计算机的其他部件的。它是怎样完成工作的呢？秘密在于软件。如果没有程序提供控制，CPU什么都不能做。程序怎样全程引导CPU工作呢？让我们从程序的基本元素——指令开始讲述。一条指令由两部分组成：一个操作码和一个或多个操作数（图1.7）。操作码告诉CPU做什么，操作数告诉CPU到哪儿去找要被操作的数据。

处理器包含四大部件（图1.8）：时钟、指令控制单元、算术逻辑运算单元（通常简称

ALU）和一组寄存器。时钟精确地生产定时电流脉冲，使之与处理器的其他部件同步。然后指令控制单元确定下一条指令的地址，并从主存中取出指令。算术逻辑运算单元执行算数操作（例如加和减）和逻辑操作（例如测试数值的真假）。寄存器是临时存储器件，它保存的是控制信息、关键数据和一些中间结果。因为寄存器在CPU上，所以它的处理速度比主存储器快。那么决定计算机速度的关键部件是什么呢？时钟！详细地说，是时钟的频率决定了计算机的处理速度。当我们购买一台计算机的时候，首先要考虑主频，即时钟的频率。

现在我们将再详细地谈一谈微处理器和中央处理器。微处理器是蚀刻在微小的硅质芯片上的中央处理单元，因此也叫微芯片。微处理器包含许多电子开关，叫作晶体管，它决定了电流是否可以通过。晶体管是微处理器的基本组成块。一个简单的芯片可能包含数万个晶体管。当电流被允许通过的时候，开关打开，这代表二进制位1。如果电流不被允许通过，开关关闭，则代表二进制位0。不同的晶体管组合代表不同的二进制组合，可以用它们来表示特定的字符、字母和数字。

Fast Reading One Input / Output System

We often mention input/output system (or I/O). What's I/O system? In computing, I/O is the communication between an information processing system (such as a computer) and the outside world. Inputs are the signals or data received by the system, and outputs are the signals or data sent out from it. I/O devices are used by a person (or other systems) to communicate with the computer. For example, the keyboard and the mouse (Pic 1.9) may be input devices for the computer, while monitors (Pic 1.10) and printers are considered as output devices. Modem (Pic 1.11) and Network Interface Cards (NIC, Pic 1.12), typically serve for both input and output devices.

Pic 1.9 Keyboard and Mouse

Pic 1.10 Monitor

Fast Reading One

Pic 1.11 Modem

Pic 1.12 NIC

The mouse and the keyboard are input physical devices. Users use them to input information, and then input devices convert it into the signal that a computer can understand. The output information from these devices is input for the computer. Similarly, the printer and the monitor take it as an input signal that a computer outputs. They convert these signals into symbols that the user can understand. These interactions between the computer and the user are called human-computer interactions.

Memory is the device which the CPU can read from and write to directly, with individual instructions. In computer architecture, the combination of the CPU and the main memory is considered as the brain of a computer, and from that point of view, any transfer of information from or to that combination, for example to or from a disk drive, is considered as I/O. The CPU and its supporting circuitry provide memory-mapped I/O that is used in low-level computer programming, such as the implementation of device drivers. An I/O algorithm is one designed to exploit locality and perform efficiently when data resides on secondary storage, such as a disk drive.

A computer uses memory-mapped I/O access hardware by reading from and writing to specific memory locations, using the same assembly language instructions that the computer would normally use to access memory.

.End.

参考译文 快速阅读1 输入/输出系统

我们常常说输入/输出系统（I/O），那什么是输入/输出系统呢？在计算机编程或应用中，输入/输出是内部信息处理系统（比如计算机）和外面世界之间的交流。输入是系统接收到的信号或是数据，输出是从系统传输出去的信号或数据。输入/输出设备是人（或其他系统）用来和计算机进行交流的设备。例如，键盘和鼠标（图1.9）对于计算机来说可能是输入设备，而显示器（图1.10）和打印机则被认为是输出设备。调制解调器（图1.11）和网卡（图1.12）既是输入设备又是输出设备。

鼠标和键盘是物理输入设备。用户使用它们输入信息，之后输入设备把这些信息转换成计算机能识别的信号。从这些设备输出的信息是计算机的输入信息。同样地，打印机和显示器把计算机的输出信号作为输入信号。它们把这些信号转换成用户可以理解的信息。这种计算机和用户之间的交流被称为人机交互。

主存是有独立指令的设备，CPU能直接对其进行读写。在计算机的架构体系中，CPU和主存的组合被认为是计算机的大脑。从这个观点出发，任何从大脑转换出的信息或到大脑的信息，例如，从大脑到磁盘驱动器或从磁盘驱动器到大脑，可认为是输入/输出（I/O）。CPU及其支持电路提供内存映射I/O，可用于低级计算机编程，如设备驱动程序的实现。一个I/O算法要在数据驻留在二级存储设备（如磁盘驱动器）时仍可开发局部性并执行指令。

使用内存映射I/O访问的计算机硬件，可阅读和写作具体的内存位置，并使用计算机通常会使用的汇编语言指令来访问内存。

Fast Reading Two History of the ENIAC

The electronic computer was one of the greatest inventions in the 20th century. Once talking about computers, we have to think of the birth of ENIAC (Electronic Numerical Integrator And Calculator) (Pic 1.13).

Fast Reading Two

Pic 1.13 ENIAC

The start of World War II produced a large need for computer capacity, especially for the military. New weapons were made so trajectory tables and other essential data were needed. In 1946, John P. Eckert, John W. Mauchly, and their associates at the Moore School of Electrical Engineering at University of Pennsylvania decided to build a high-speed electronic computer to do the job. This machine became known as ENIAC.

The size of ENIAC's numerical "word" was 10 decimal digits, and it could multiply two of these numbers at a rate of 300 per second, by finding the value of each product from a multiplication table stored in its memory. ENIAC was therefore about 1,000 times faster than the previous generation of relay computers.

ENIAC used 18,000 vacuum tubes, about 1,800 square feet of floor space, weighed 30 tons and consumed about 180,000 watts of electrical power. It had punched card I/O, 1 multiplier, 1 divider/square rooter, and 20 adders using decimal ring counters, which served as adders and also as quick-access (0.0002 seconds) to read-write register storage. The executable instructions making up a program were embodied in the separate "units" of ENIAC, which were plugged together to form a "route" for the flow of information.

ENIAC was commonly accepted as the first successful high-speed electronic digital computer (EDC) and had been used from 1946 to 1955, but it had a number of shortcomings which were not solved, notably the inability to store a program. A number of improvements were also made to ENIAC from 1948, based on the ideas of the Hungarian-American mathematician, John Von Neumann (Pic 1.14) who was called the father of computer.

John Von Neumann contributed a new awareness of how practical, yet fast computers should be organized and built. These ideas, usually referred to as the stored-program technique, became essential for future generations of high-speed digital computers and were universally adopted. Electronic Discrete Variable Automatic Computer (EDVAC) designed by John Von Neumann was built in 1952. This computer used 2,300 vacuum tubes, but its speed was 10 times faster than ENIAC which used 18,000 vacuum tubes. And the most importantly, Random Access Memory (RAM) was used.

Pic 1.14 John Von Neumann

.End.

参考译文 快速阅读2 ENIAC的历史

电子计算机是20世纪最伟大的发明之一。一提起计算机，我们就不得不想到ENIAC（电子数字积分器和计算器）（图1.13）的问世。

第二次世界大战的爆发对计算机的能力提出了更高的要求，尤其是在军事领域。新武器的制造需要弹道表和其他关键数据。1946年，John P.Eckert、John W.Mauchly和他们在宾州大学摩尔电器工程学院的同事决定制造一台高速电子计算机来完成这项工作。这台机器被称为ENIAC（埃尼亚克）。

ENIAC的数字“字长”为10位十进制数字，它能以每秒300次的速度运算两个这样数字的乘法，其方法是每次从存储器中的乘法表中找到乘积的值。因此，ENIAC具有比它的前一代继电器计算机快约1000倍的速度。

ENIAC使用了18,000个电子管，占地约1,800平方英尺，重达30吨，消耗大约180,000W电能。它有穿孔卡片I/O（输入/输出设备）、一个乘法器、一个除法器/平方根器和使用十进制循环计数器的20个加法器，这些设备既可用作加法器，又可快速访问（0.0002秒）读/写寄存存储器。在ENIAC里，可执行的指令组成一个程序，包含于单独的“单元”里，这些单元连接起来为信息流形成一个“路由”。

ENIAC被普遍认为是第一个成功的高速电子数字计算机，并在1946年至1955年中得到应用。但是它的许多缺点没有得到解决，尤其是不能存储程序。从1948年起，ENIAC有了许多改进，其思想来源于被人们称为“计算机之父”的美籍匈牙利数学家约翰·冯·诺依曼（图1.14）。

约翰·冯·诺依曼提出了如何组建一个应用型的、快速运算的计算机的新见解。这些思想（通常指存储技术）对后续几代高速数字计算机的发展十分必要，因此被普遍采纳。1952年，由约翰·冯·诺依曼设计的电子计算机EDVAC问世。这台计算机共使用了2,300个电子管，运算速度却比拥有18,000个电子管的“埃尼亚克”提高了10倍。最重要的是，随机存取存储器（RAM）被采用了。

Ex 1 **What is a computer like in your mind? Try to give a brief summary of this passage in no more than five sentences.**

Ex 2 **Fill in the table below by translating the Chinese or English.**

display screen	
	打印机
scanner	
	内存
output device	
	主板
LCD	
	随机存取存储器
digital camera	
	文件夹
pointing device	
	传真机

Ex 3 **Choose the best answer for each of the following statements according to the text we've learnt.**

1. A computer system consists of two main elements: ________.
   A. input devices, output devices B. hardware, software
   C. CPU, memory D. RAM, ROM
2. The peripherals of the computer can be classified into two groups: ________.
   A. Data Bus, cables
   B. pointing devices, output devices
   C. input devices, output devices
   D. Control Bus, inner Bus
3. Users can type in texts and commands using the ________, which is one of the most important input devices.
   A. mouse B. scanner
   C. microphone D. keyboard

4. The most useful pointing device is a ________, which allows the user to point to elements on the screen.

A. digital camera　　B. mouse

C. PC camera　　D. keyboard

5. The characteristic of the LCD is ________.

A. taking up a lot of space

B. taking up little space

C. displaying texts and images with greater unclarity

D. A and C

6. ________ is the control and data processing center of the whole computer system.

A. Memory　　B. Motherboard

C. CPU　　D. Bus

7. When the computer is turned off, any information in ________ will be lost.

A. RAM　　B. ROM

C. storage　　D. memory

8. ________ is the most popular Bus to be used on a motherboard.

A. Address Bus　　B. Data Bus

C. Control Bus　　D. PCI Bus

9. Two main components on the motherboard are ________.

A. memory and circuit boards

B. CPU and memory

C. arithmetic unit and control unit

D. ROM and RAM

10. A ________ does not belong to the output devices.

A. facsimile machine　　B. headset

C. speaker　　D. microphone

Part B Practical Learning

Task One

Task Two

Training Target

In this part, there are two tasks in English environment. You should complete these tasks in groups under the joint guidance of professional teachers and laboratory teachers, so as to train and improve your ability to complete professional tasks in English environment.

Task One Name the Computer Hardware Devices

The first task is to name the computer hardware devices. In this task, students must know the composition of the computer hardware system (Pic 1.15 and Pic 1.16), and their English names.

There is some information about the hardware devices of a computer. The information can help students finish the task.

As we know, the computer hardware system usually consists of Calculator, Controller, Memory and I/O . Calculator and Controller are often referred to CPU. Memory includes the main memory and the secondary storage (such as hard disk). The basic input devices include keyboard and mouse; the basic output devices include display screen, printer and so on. These devices must be plugged in the motherboard through the sockets or slots (Pic 1.17).

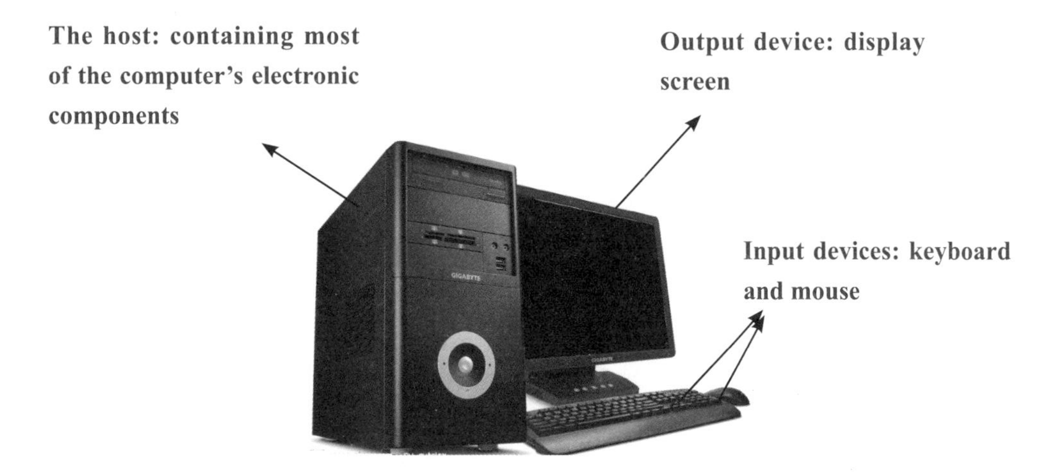

Pic 1.15 The Computer

Pic 1.16 The host

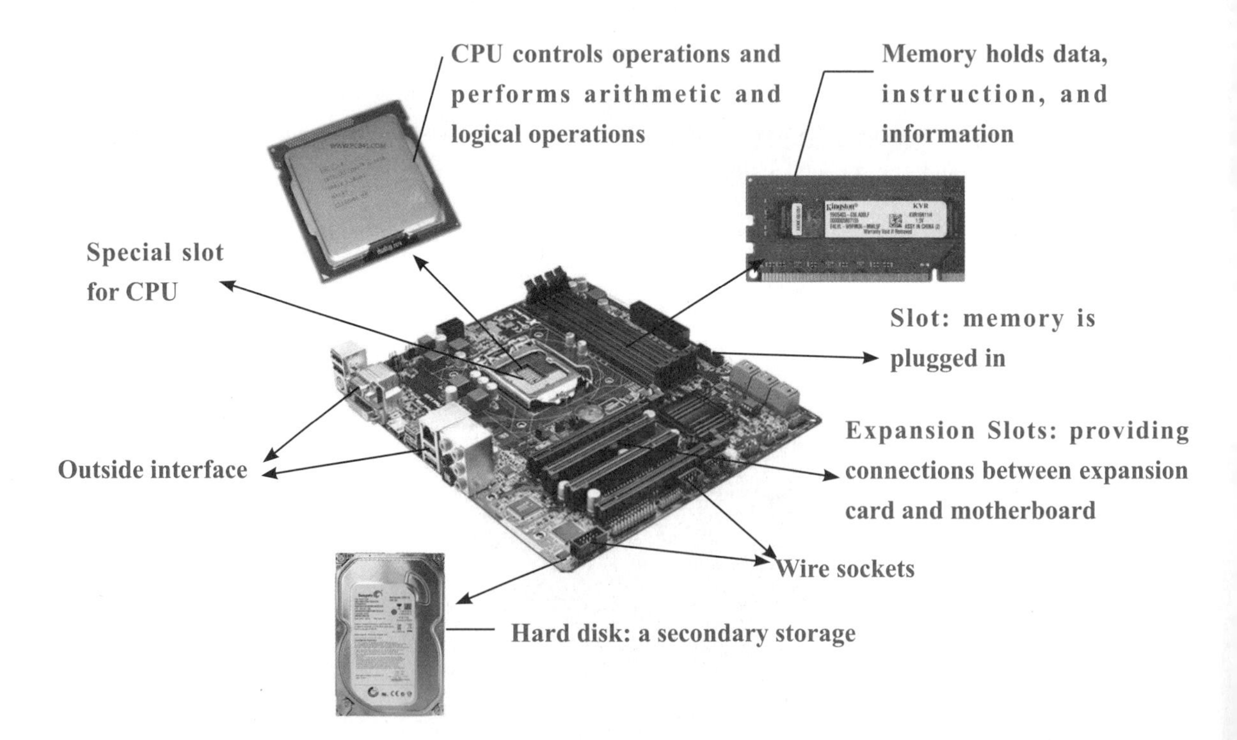

Pic 1.17 The motherboard

Task Two Computer Assembly and CMOS Setup

In this task, firstly, students can assemble the computer. Secondly, students can set up the CMOS after finishing the assembling.

There is some information about the CMOS.

Complementary metal-oxide-semiconductor (CMOS) on the motherboard records the date, time, hard disk references, and other advanced references about computer. Knowing how to access and change settings in your BIOS or CMOS can save you a lot of headache when troubleshooting a computer.

First: How to enter the CMOS?

There are numerous ways to enter the CMOS setup. Below is a list of the majority of these methods as well as other recommendations for entering the CMOS setup.

As the manufacturer's logo appears, press the designated setup button to enter the CMOS. The keys are decided by the manufacturer. Typical setup keys are F2, F10, F12, ESC and DEL. The keys will be displayed on the screen with the manufacturer's logo (Pic 1.18).

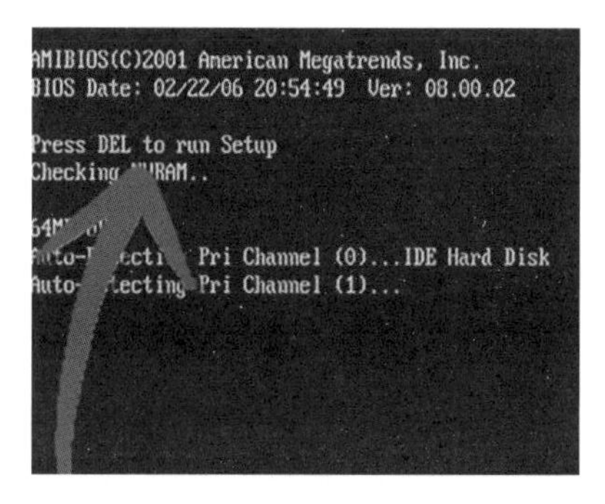

Pic 1.18 The setup key

If you strike the DEL key immediately, you will enter the CMOS setup (Pic 1.19).

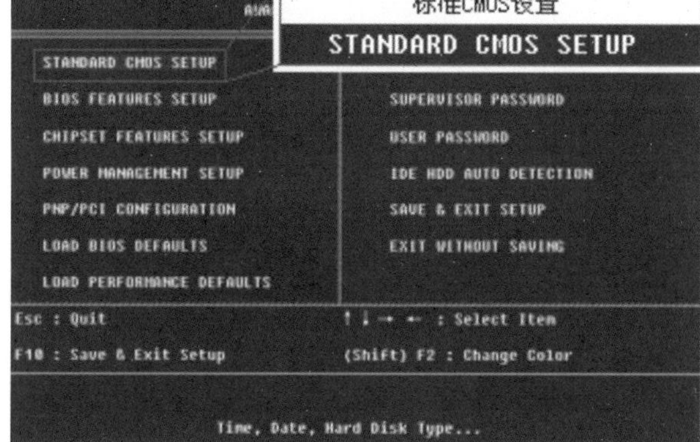

Pic 1.19 CMOS setup

The details of CMOS setup (Pic 1.20) are shown below.

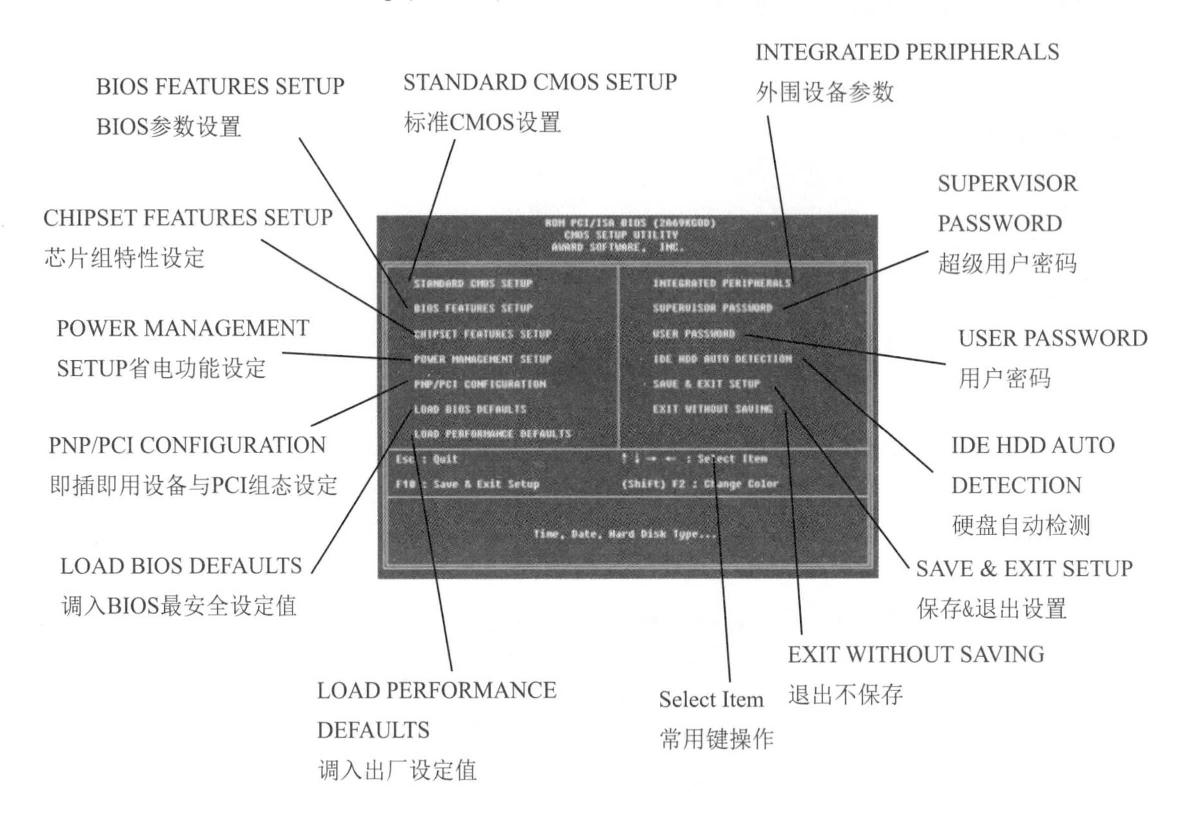

Pic 1.20 The details of CMOS setup

Part C Occupation English

Occupation English

In this part, there is an English dialogue in real life and work environment. You will play the roles of A and B and read the dialogue aloud to practice your ability to use English.

An Introduction to Computers for Customers
给客户进行电脑介绍
Character Setting: a Computer Salesman (A) and a Customer (B)
角色设置：计算机销售人员（A）和客户（B）

A: Hello, welcome to Huawei mall. Can I help you?您好，欢迎光临华为商城。需要帮忙吗？

B: Yes, I need to buy a computer.是的，我需要买一台电脑。

A: OK, can you tell me what kind of computer you need? Because we have several types of computers, such as Huiwei MateBook Series, MagicBook Series, MagicBook Pro Series and so on.好的，那您能告诉我您需要一台什么类型的电脑吗？因为我们有好几种类型的电脑，像Huiwei MateBook系列，MagicBook 系列，MagicBook Pro系列等。

B: I want a computer with better functions. I have roughly browsed the relevant information of Huawei computers on the Internet. I like one of the computers — MateBook 14. Can you give me a specific introduction?我想要一台功能强大的电脑。我在网上大概浏览了一下华为电脑的相关资料，我对其中一款电脑 —— MateBook 14比较中意。您能给我具体介绍下吗？

A: OK, you have a good eye. This is a powerful computer. This computer has a brand new 2K multi-touch full screen, which is one of its highlights. And its configuration is extremely high: it carries the 10th generation Intel Core i7-1010510U processor, which can reach up to 4.9GHz; it is equipped with 16GB memory and 512GB solid-state hard disk, which are large enough; in addition, it carries a NVIDIA GeForce MX350 independent graphics card, which makes video editing and game experience smoother and faster. There are also 2 * 2 MIMO dual antenna WLANs. The theoretical transmission rate can reach 1 733Mbps, and the connection is more stable. If you buy this computer, it will be enough for daily work, surfing the Internet, watching videos, playing games, etc.好的，您的眼光非常好，这是一款功能强大的电脑。这款电脑拥有全新配备的2K全面屏，多点触控，这是它的一大亮点。而且它的配置极高：它搭载的是第十代Intel酷睿i7-1010510U处理器，睿频至高可达4.9GHz；配有16GB内存、512GB固态硬盘，容量足够大；另外它还搭载了NVIDIA GeForce MX350独立显卡，无论是视频编辑，还是游戏体验都更加流畅、快速。还有2*2MIMO双天线WLAN，理论传输速率至高可达到1,733Mbps, 连接更稳定。您如果买了这款电脑，对于日常办公、上网冲浪、看视频、打游戏等都足够用了。

B: It sounds like a great computer. If I buy it today, do you have any discount? 听起来这款电脑很不错。如果我今天买，你们店有什么优惠活动吗？

A: Well, it's a coincidence that you came here. Our store is really offering a discount. If you buy this computer, you can get a 10% discount. 嗯，您来得实在是太巧了，我们确实在实行优惠活动。如果您购买这款电脑，现在可以打九折优惠。

B: Ten percent discount for this computer?这款电脑打九折？

A: Of course, you've heard correctly. If you buy this computer today, you will get a 10% discount. This is an activity that has never been done before. You are very lucky. Do you think you'll buy it today? It's really suitable!是的，您没有听错。您今天购买的话，这款电脑打九折。这是以前从没有过的活动，您很幸运。您看您今天买吗？真的很合适！

B: Wow, I am so lucky to catch up with Huawei MateBook 14 discount. I have been paying attention to it for a long time!哇，我真是太幸运了，居然赶上了华为MateBook 14 笔记本打折，我关注它好久了！

A: That's right. What are you waiting for? Order it! 对呀，那您还等什么，下单吧！

B: Of course. Order and buy it!那当然了，下单吧，就买它了！

Word Building

前缀/后缀由一个或几个字母组成，放在词根或单词之前/之后，组成一个新词。

（1）auto-（前缀）：自动的

alarm 报警器 ———— autoalarm 自动报警器

code 编码 ———— autocode 自动编码

（2）co-（前缀）：共同

action 行动 ———— coaction 共同行动

operate 操作 ———— cooperate 协作，合作

(3) fore-(前缀):预……,前……

head 头 ———— forehead 前额

word 话 ———— foreword 前言,序言

(4) -ess(后缀):女性的

mayor 市长 ———— mayoress 女市长

actor 男演员 ———— actress 女演员

(5) -ee(后缀):被……的人,受……的人

employ 雇用———— employee 雇员

test 测试———— testee 被测验者

Ex Try your best to guess the meaning of each word on the right according to the clues given on the left.

correction	校正(名词)	autocorrection	______
biography	传记(名词)	autobiography	______
exist	存在(动词)	coexist	______
author	作家(名词)	coauthor	______
tell	告诉(动词)	foretell	______
father	父亲(名词)	forefather	______
waiter	男服务员(名词)	waitress	______
host	主人(名词)	hostess	______
train	训练(动词)	trainee	______
pay	薪水(名词)	payee	______

Ex 1 When we use a computer, what does the processor do?

Ex 2 Fill in the table below by translating the Chinese or English.

processor	
	插座
microprocessor	
	寄存器
operation code	
	开关
main frequency	
	脉冲
operand	

Ex 3 Choose the best answer for each question or statement according to the text we've learnt.

1. Which of the following is not one of the four parts that comprise a processor? ________

   A. clock. B. memory.

   C. register. D. ALU.

2. A computer system contains input device, output device, CPU and ________.

   A. ROM B. RAM

   C. memory D. register

3. The processor consists of ________.

   A. clock, ALU, register, and operation code

   B. ALU, instruction control, clock, and register

   C. CPU, memory, register, and clock

   D. instruction control, clock, operation code, and operand

4. ________ is the soul and heart of a computer, and it can manipulate data and control the rest part of the computer.

   A. Memory B. Input device

   C. Register D. CPU

5. Which of the following tells the CPU what to do? ________

   A. instruction. B. operand.

   C. operation code. D. arithmetic and logic unit.

6. Which of the following tells the CPU where to find the data to be manipulated? ________

   A. clock. B. instruction control unit.

   C. operation code. D. operand.

7. When we buy a computer, we usually consider the ________ first, and that means a clock's frequency.

   A. main frequency B. board

   C. display D. keyboard

8. The arithmetic unit performs the ________ operations.

   A. arithmetic B. comparison

   C. logical D. all of the above.

9. ________ holds control information, key data and some intermediate results.

   A. Register B. Instruction control unit

   C. ALU D. Storage

10. When electric current is not allowed to pass through transistors on the microchip, the switch is ________ and it represents a ________ bit.

    A. off, 1 B. on, 1

    C. on, 0 D. off, 0

Project Two

Installing the Software

Part A Theoretical Learning

Part B Practical Learning

Part C Occupation English

Part A Theoretical Learning

Training Target

In this part, our target is to improve the speed of reading professional articles and the comprehension ability of the reader. We have marked specialized key words and some flexible sentences. Try to grasp the main idea of each paragraph.

Skill One

Skill One System Software — OS

System software, or systems software, is the computer software designed to provide a **platform** to other software. Examples of system software include **operating system**, scientific computational software, game engines, industrial automation, and software as service applications. In contrast to system software, software that allows users to do things like creating text documents, laying games, listening to music, or surfing the Web is called **application software**.

system software系统软件
software ['sɒftweə(r)] *n.* 软件
platform ['plætfɔ:m] *n.* 平台
operating system 操作系统
application software 应用软件

Operating System or System Control Programs

An operating system scientific is the system software that manages computer hardware and software resources and provides common services for computer programs. It provides a platform (hardware abstraction layer) to run high-level system software and application software. A **kernel** is the core part of the operating system that defines an **API** for application programs (including some system software) and an interface to device drivers.

kernel ['kɜ:n(ə)l] *n.* 核心；要点
API=Application Program Interface 应用程序接口

Examples of Operating System

Microsoft Windows

Microsoft Windows is a family of **proprietary** operating systems designed by Microsoft Corporation and primarily targeted to Intel **architecture**-based computers, with estimated 88.9 percent total usage share on Web-connected computers. The latest **version** is Windows 10.

In 2011, Windows 7 overtook Windows XP as the most common version in use. Microsoft Windows was first **released** in 1985, as an operating environment running on top of MS-DOS, which was the

proprietary [prə'praɪətri] *adj.* 专有的，专利的
architecture ['ɑ:kɪtektʃə(r)] *n.* 结构，体系结构
version ['vɜ:ʃn] *n.* 版本，译文
release [rɪ'li:s] *vt.* 发布

standard operating system shipped on most Intel architecture personal computers at the time. In 1995, Windows 95 was released which only used MS-DOS as a **bootstrap**.

bootstrap ['bu:tstræp]
n. 引导程序

For backwards compatibility, Win9x could run real-mode MS-DOS and 16-bit Windows 3.x drivers. Windows ME, released in 2000, was the last version in the Win9x family. Later versions have all been based on the Windows NT kernel. Current client versions of Windows run on IA-32, x86-64 and 32-bit ARM microprocessors.

Server editions of Windows are widely used. In recent years, Microsoft has expended significant capital in an effort to promote the use of Windows as a server operating system.

Unix and Unix-like Operating Systems

Unix was originally written in assembly language. Ken Thompson wrote B, mainly based on BCPL, and its experience in the MULTICS project. B was replaced by C, and Unix, rewritten in C, developed into a large, complex family of inter-related operating systems which have been **influential** in every modern operating system.

influential [ˌɪnflu'enʃl]
adj. 有影响的

The Unix-like family is a **diverse** group of operating systems, with several major sub-categories including System V, BSD, and Linux. The name "UNIX" is a trademark of The Open Group which **licenses** it for use with any operating system that has been shown to conform to their definitions. Unix-like is commonly used to refer to the large set of operating systems which resemble the original UNIX.

diverse [daɪ'vɜ:s]
adj. 不同的
license ['laɪsns]
v. 批准，许可

Linux

The Linux kernel originated in 1991, as a project of Linus Torvalds, who was a university student in Finland. He posted information about his project on a newsgroup for computer students and programmers, and received support and assistance from volunteers who succeeded [in creating a complete and functional kernel.]

volunteer [ˌvɒlən'tɪə(r)]
n. 义务工作者

Linux is Unix-like, but was developed without any Unix code, unlike BSD and its **variants**. Because of its open license model, the Linux kernel code is available for study and modification, which resulted in its use on a wide range of computing machinery from supercomputers to smart-watches. Although estimates suggest that Linux is used on only 1.82% of all "desktop" (or laptop) PCs, it has been widely adopted for use in servers and embedded systems like cellp hones.

variant ['veəriənt]
n. 变体

Mac OS

Mac OS (formerly “Mac OS X” and later “OS X”) is a line of open core graphical operating systems developed, marketed, and sold by Apple Inc., the latest of which is pre-loaded on all currently shipping Macintosh computers. Mac OS is the successor to the original classic Mac OS, which had been Apple’s primary operating system since 1984. Mac OS is a UNIX operating system built on technology that had been developed at NeXT through the second half of the 1980s and up until Apple purchased the company in early 1997. The operating system was first released in 1999 as Mac OS X Server 1.0, followed in March 2001 by a client version. Since then, six more distinct “client” and “server” editions of Mac OS have been released, until the two were merged in OS X 10.7 “Lion”.

.End.

Key Words

system software 系统软件
platform *n.* 平台
proprietary *adj.* 专有的，专利的
architecture *n.* 结构，体系结构
bootstrap *n.* 引导程序
diverse *adj.* 不同的
volunteer *n.* 义务工作者
operating system 操作系统
API 应用程序接口
software *n.* 软件
kernel *n.* 核心；要点
version *n.* 版本，译文
release *vt.* 发布
influential *adj.* 有影响的
license *v.* 批准，许可
variant *n.* 变体
application software 应用软件

参考译文 技能1 系统软件——操作系统

系统软件，是为其他软件提供平台的计算机软件。系统软件包括操作系统、科学计算软件、游戏引擎、工业自动化和服务应用软件。与系统软件相反，允许用户做诸如创建文档、设置游戏、听音乐或上网冲浪的软件被称为应用软件。

操作系统或系统控制程序

操作系统（OS）是管理计算机硬件和软件资源并为计算机程序提供公共服务的系统软件。它为运行高级系统软件和应用软件提供了一个平台（硬件抽象层）。内核是操作系统的核心部分，它定义了应用程序（包括一些系统软件）的接口和设备驱动程序的接口。

操作系统的示例

Microsoft Windows系统

Microsoft Windows是一个专有操作系统的家族，它由微软公司设计，主要针对以英特尔公司体系结构为基础的计算机，估计使用量占共享网络连接计算机总量的88.9%。最新版本是Windows 10。

2011年，Windows 7取代了Windows XP，成为最常用的版本。Microsoft Windows系统于1985年首次被发布，作为一种操作环境运行在MS-DOS之上的系统，它是当时大多数英特尔架构个人计算机中的标准操作系统。1995年，Windows 95被发布，只使用MS-DOS作为引导程序。

为了向后兼容，Win9x可以运行实模MS-DOS和16位Windows 3.x驱动程序。2000年发布的Windows ME是Win9x家族中的最后一个版本。后来的版本都基于Windows NT内核。当前Windows客户端版本在IA-32，x86-64和32位ARM微处理器上运行。

Windows的服务器版本被广泛使用。近几年来，微软花费了大量资金来推广Windows作为服务器操作系统的使用。

UNIX和类UNIX操作系统

UNIX最初是用汇编语言编写的。Ken Thompson写的B主要基于BCPL和它在MULTICS项目中的经验。B被C取代，UNIX被重写成C，后来UNIX发展成一个大型、复杂的相互关联的在每个现代操作系统中都有影响力的操作系统家族。

类UNIX的家庭是一组不同的操作系统。几个主要的子类别有System V、BSD和Linux。“UNIX”的名称是开放组的一个标志，它为那些使用显示符合它们定义的操作系统颁发许可证。类UNIX通常用于指一大套与原始UNIX类似的操作系统。

Linux

Linux内核起源于1991年，是Linus Torvalds在芬兰还是大学生时候的一个项目。他面向计算机学生和程序员发布了有关这个项目的信息，并得到了志愿者的支持和帮助，并成功创建了一个完整的功能内核。

Linux是类UNIX的，但它的开发没有任何的UNIX代码，不像BSD和它的变体。因为它有开放许可证模型，Linux内核代码可用于学习和修改，这使得它的使用范围很广从超级计算机到智能手表。尽管预计Linux在台式机(或笔记本电脑)的使用仅占1.82%，但它已被广泛用于服务器和嵌入式系统，如手机。

Mac OS

Mac OS(以前的“Mac OS X”和后来的“OS X”)是由苹果公司开发、推广和销售一套开放核心图形操作系统，最新的一款产品已预装在目前所有出货的麦金塔电脑中。Mac OS是最初的经典Mac OS的继承者，后者自1984年以来一直是苹果的主要操作系统。与它的前身不同的是，Mac OS是一种在20世纪80年代后半段开发出来的基于NeXT技术UNIX操作系统，直到1997年初苹果公司收购了该公司。该操作系统于1999年首次作为Mac OS X服务器1.0被发布，接着在2001年3月作为客户端版本(Mac OS X v10.0“Cheetah”)被发布。从那以后，六个以上的Mac OS的“客户端”和“服务器”版本已经被发布，直到这两种版本被合并到OS X 10.7“Lion”中。

Skill Two

Skill Two Application Software — OA

Application software helps you accomplish specific tasks. You can use application software to write letters, manage your **finance**, draw pictures, play games and so on. Application software is also called software, an application or a program. You can buy software at computer stores. There are also thousands of programs available on the Internet.

finance ['faɪnæns] *n.* 财务

Software you buy at a computer store is usually on a single **CD-ROM** disk, a **DVD-ROM** or several **floppy disks**. Before you use the software, you install, or copy the content of the disk or disks to your computer. Using a CD-ROM or DVD-ROM disk is a fast method of **installing** software.

CD-ROM 光盘只读存储器
DVD-ROM 数字化视频光盘存储器
floppy disk 软盘
install [ɪn'stɔ:l] *v.* 安装

When a manufacturer adds new **features** to the existing software, the updated software is given a new name or a new version number. This helps people distinguish new versions of the software from older versions. Manufacturers may also create minor software updates, called patches, or improvements to software. A patch is also often referred to as a service pack.

feature ['fi:tʃə(r)] *n.* 属性

Bundled software is the software that comes with a new computer system or device, such as a printer. Companies often provide bundled software to let you start using the new equipment right away. For example, new computer systems usually come with word processing **spreadsheet** and **graphics** programs. Most softwares come with a built-in help feature and printed **documentation** to help you learn to use the software. You can also buy computer books that contain detailed, step-by-step instructions or visit the manufacturer's **web site** for more information about the software. The **OA** is the technology that reduces the amount of human effort necessary to perform tasks in the office.

bundle ['bʌndl] *v.* 绑定
spreadsheet ['spredʃi:t] *n.* 电子表格
graphics ['græfɪks] *n.* 绘图
documentation [ˌdɒkjumen'teɪʃn] *n.* 文件
web site 网站
OA office automation 办公自动化

Today's businesses have a wide variety of OA technology at their **disposal**, such as data processing, **word processing, graphic** processing, image processing, voice processing and networking. The widespread use of the OA technology began in the workplace of offices, banks and factories, etc. The development of OA system has been synchronizing with the development of data, information and

disposal [dɪ'spəʊzl] *n.* 处理
word processing 文字处理
graphic ['græfɪk] *n.* 图形

knowledge. The first OA system is mainly used to deal with data, the second OA system is chiefly used to deal with information, and the third generation is used to deal with knowledge. These three stages in the development have accomplished the **leap** from the data processing to the information processing, as well as the leap from the information processing to the knowledge processing. In the development of OA system, its scope for using increasingly widened, the dealings with the content promoted step by step, and the system function was perfect ultimately.

The new version of Office suite has been publicly available from Microsoft. The new version of Office suite has build-in native **XML** support for Excel and Access. Excel users will find new features for working with web components so that the used Excel might **automatically** publish a **web page**, or a **pivot** table/chart. **Smart Tags** will appear on the screen while you work, offering information on completing tasks faster. For example, Smart Tags will display **AutoCorrect, AutoFormat** and **Paste** options that can keep users from searching through **menus**. There are also many features found in the new version of Office designed to help when data losing has occurred. Excel, PowerPoint, and Outlook will join Word in offering AutoRecover feature, which automatically saves documents and data at **intervals**.

Today's organizations have a variety of OA hardware and software components at their disposal. The list includes telephone computer systems, electronic mail, word processing, **desktop publishing**, database management system, two-way cable TV, office-to-office satellite broadcasting, online database service, and voice recognition and synthesis. Each of these components is intended to automate a task or function that is presently performed manually.

.End.

leap [li:p] *n.* 飞跃

XML(Extensible Markup Language) 可扩展标记语言

automatically [ˌɔ:tə'mætɪkli] *adv.* 自动地

web page 网页

pivot ['pɪvət] *adj.* 重要的

Smart Tag 智能标记

AutoCorrect 自动纠错

AutoFormat 自动格式化

paste [peɪst] *vt.* 粘贴

menu ['menju:] *n.* 菜单

interval ['ɪntəvl] *n.* 间隔

desktop publishing 桌面印刷

Key Words

finance *n.* 财务	floppy disk 软盘
install *v.* 安装	feature *n.* 属性
bundle *v.* 绑定	spreadsheet *n.* 电子表格
graphics *n.* 绘图	documentation *n.* 文件
web site 网站	disposal *n.* 处理
graphic *n.* 图形	leap *n.* 飞跃
automatically *adv.* 自动地	word processing 文字处理
web page 网页	pivot *adj.* 重要的
smart Tag 智能标记	AutoCorrect 自动纠错
AutoFormat 自动格式化	past *vt.* 粘贴
menu *n.* 菜单	interval *n.* 间隔
CD-ROM 光盘只读存储器	DVD-ROM 数字化视频光盘存储器
OA office automation 办公自动化	XML 可扩展标记语言
desktop publishing 桌面印刷	

参考译文 技能2 应用软件——办公自动化

应用软件帮助你完成特定的任务。你可以用应用软件写信、管理财务、画画、玩游戏等。应用软件也叫软件、应用程序或程序。你可以从计算机商店中买到软件，还可以从互联网上获取成千上万个软件。

你从计算机商店中购买的软件通常存储在简单的CD盘、DVD盘或者几张软盘中。在你使用这些软件之前，需要安装或者将这些盘上的内容拷贝到计算机上。使用CD盘或者DVD盘是一种较快的安装软件的方式。

当制造商向已经存在的软件上面添加新的属性时，通常会给它起一个新的名字或者是新的版本号。这有助于人们把新版本的软件与旧版本的软件区分开来。制造商也可以做一些小规模的软件升级，叫作补丁，也叫升级软件。补丁也经常被称作服务包。

绑定软件是指那些与一些新的计算机系统或者设备（例如打印机）一起售出的软件。公司通常会提供绑定软件，以便让你马上开始使用新设备。例如，新的计算机系统通常会附带一些文字处理软件、电子表格软件以及绘图软件。大多数软件带有一个内置的帮助功能和打印文档来帮助你学习使用软件。你也可以购买包含详细信息的电脑书，一步一步地来学习使用软件，或者访问制造商的网站以获取软件更多的信息。办公自动化技术可以减少人们在办公室里执行任务时所花费的人力。

如今的企业有各种办公自动化技术可用，如数据处理、文字处理、图形处理、图像处理、声音处理和网络。办公自动化技术的广泛应用始于办公室、银行和工厂等办公场所。办公自动化系统的发展和数据、信息和知识的发展同步。第一代办公自动化系统主要用于处理数据；第二代办公自动化系统主要用于处理信息；而第三代则用于处理知识。这三个阶段的发展已经完成了从数据处理到信息处理，以及从信息处理到知识处理的飞跃。随着办公自动化系统的发展,其使用范围日益扩大，处理内容逐步提升，系统功能日趋完善。

微软已经推出了新版Office，新版Office为Excel和Access内建了XML支持。Excel用户在使用web组件时会发现新功能，即可以自动地发布网页或重要图标。用户工作时智能标记会出现在屏幕上，提供信息并使任务完成得更快。例如，智能标记显示自动纠错、自动格式化以及粘贴项，从而使用户不必去搜寻菜单。当产生数据丢失后，新版Office中会有许多新功能帮助处理相关问题。Excel、PowerPoint和Outlook将同Word一样提供自动恢复功能，每间隔一段时间便会自动存储文档和数据。

如今的机构已经配置了各种各样的办公自动化硬件和软件，包括电话、计算机系统、电子邮件、文字处理、桌面印刷系统、数据库管理系统、双向电缆电视、办公室对办公室的卫星广播、联机数据库服务、声音识别及合成系统。这些配备都力图使目前需手工完成的任务或功能自动化。

Fast Reading One Computer Development

Fast Reading One

Wearable computers are the next wave of portable computing and they will go way beyond laptops.

What's a Wearable Computer?

Imagine watching a movie projected through your eyeglasses onto a virtual screen that seems to float in your field of vision. Or imagine working in an automobile, an airplane, or on an underwater mission and reading an instruction manual, communicating with co-workers, or inputting data via computer — all without lifting a finger from the task at hand.

To date, personal computers have not lived up to their name. Most machines sit on the desk and interact with their owners for only a small fraction of the day. Smaller and faster notebook computers have made mobility less of an issue, but the same staid user paradigm persists.

A wearable computer hopes to shatter this myth of how a computer should be used. A wearable computer is a very personal computer. A person's computer should be worn, such as eyeglasses or clothing are worn, and interacted with the user based on the context of the situation. With heads-up displays, unobtrusive input devices, personal wireless local area networks, and a host of other context sensing and communication tools, the wearable computer can act as an intelligent assistant.

Micro Optical Corp., based in Westwood, Mass., has designed two models of an eyeglass display — one that clips onto the side of the user's glasses, and the other integrated directly into the eyewear. Though the user still needs a CPU — a laptop, a wearable computer, or even a DVD player or cell phone — the monitor or screen is actually projected through the user's eyeglasses.

Gerg Jenkins, sales manager at Micro Optical, says the LCD is positioned near the user's temple. A projected image passes through the lenses of regular eyeglasses, bounces off a mirror, and displays the illusion of a full-size monitor floating in front of the user's face. The display weighs less than an ounce, so it's much more comfortable than some of the earlier head-mounted displays.

WetPC

Called the WetPC[①] (Pic 2.1), it comprised a miniature personal computer with a mask-mounted virtual display and a novel one-handed controller — called a Kord Pad[②]. The computer was mounted in a waterproof housing on the diver's air tank. A cable from it was connected to the waterproof virtual display which presented the diver with a high contrast display "floating" in the field of view. A second cable was connected to the Kord Pad, a 5-key device which the diver could hold in either hand and was used to control the computer by pressing single or multiple keys. A Graphical User Interface (GUI) showed the user which key (or keys) to press. The GUI facilitated the wear ability and usability of the WetPC — underwater computer. It was the result of several years of research into interface design and functionality. Rather like playing the piano, the user can interact with the computer in a very natural way — the diver can access and record information with one hand, even while swimming.

Pic 2.1 WetPC

The WetPC can help salvage-divers, maritime archeologists, and police divers find objects, record or look for information, or simply monitor their locations at all times. Scientists can use the unit for mapping and monitoring coral reefs. Navy divers can use the WetPC to search for mines and other unexplored devices.

The WetPC has applications in education, recreation, and tourism as well. Divers can use the WetPC to navigate reefs and create a digital-guided tour for underwater tourists.

The Challenges of a Wearable Computer

The creation of a wearable computer is not without challenge — primarily, how to produce an affordable product. Most units aren't priced for the average consumers. On the development side, designers and engineers continually strive to increase the functionality and comfort of wearable computers.

"One of the big challenges for us is to continue to make them user-friendly — to maintain usability as the functionality goes up and the size goes down." says Danny Cunagin, president of Logic.

And most wearable computers use standard desktop computer specifications — Pentium processors, Windows operating systems, creating a user interface that is easy, unobtrusive, and comfortable to use.

.End.

① WetPC:湿PC，一种可在水下使用的计算机。

② Kord Pad: 五键考德机，一种计算机输入设备。

参考译文 快速阅读1 计算机的发展

可佩带式计算机是便携式计算机技术的下一个浪潮，而且将远远超过笔记本电脑。

什么是可佩带式计算机？

想象一下，将电影通过你的眼镜投影到虚拟屏幕上的情形，好像电影画面就浮现在你的眼前！或者再想象一下，你正在开车、开飞机或执行海底任务时，读取操作指令，和同事进行交流，或者把数据输入到计算机中——所有的这一切都没有动一下手指头。

时至今日，个人计算机已无法名副其实了。大多数计算机被放置在桌子上，只在一天中很短的时间内被它们的主人使用。体积更小、速度更快的笔记本使便携性不成问题，但是固定的用户模式依然未变。

可佩带式计算机有望突破这种使用模式。可佩带式计算机是绝对的个人计算机。一台个人计算机应该被随身佩带，就像眼镜或衣服一样。根据环境的不同，它可以与用户进行相应方式的互动。可佩带式计算机具有平视显示器、不引人注意的输入设备、个人无线局域网络、许多相关的感知和交流工具，可以充当一个智能助手。

位于美国马萨诸塞州韦斯特伍德的Micro Optical公司已经设计出两种基本的眼镜显示器模型：一种夹在用户的眼镜边上，另一种直接与眼镜结合在一起。尽管用户仍需要一个CPU，可以是笔记本、可佩带式计算机，甚至可以是DVD播放器或手机，但它们的显示器或屏幕已经可以通过用户的眼镜进行投影了。

Micro Optical公司的销售经理Gerg Jenkins说液晶显示器被放置在用户的太阳穴附近。通过眼镜镜片投影的图像从镜片上弹起，在用户的面前浮现出全屏的幻觉似的影像。这个显示器的重量不足一盎司，因此它比以前那种戴在头顶上的显示器舒服多了。

湿PC

所谓湿PC（图2.1），是一个小型的个人计算机，包含一个嵌在面罩内的显示器和一个独特的可由一只手操作的控制器（五键考德机）。这个计算机嵌在潜水员防水的氧气罐里。计算机的一条电缆连接到防水的显示器上，在潜水员眼前形成一个高对比度的“漂浮”画面；另一条电缆连接到五键考德机——一个只有五个键的设备上。潜水员可以用任意一只手拿着，并且可以通过按一个或多个键来控制计算机。图形用户界面告诉用户该按哪一个或哪几个键。图形用户界面的应用增强了湿PC的水下可佩带性和可用性，这是几年来对界面设计和功能性研究的结果。就像弹钢琴一样，用户可以用一种非常自然的方式操作计算机，即使潜水员正在游泳，也可以用一只手存取和记录信息。

湿PC能够帮助海上救助潜水员、海上考古学家和海上警察寻找目标，记录或查找信息，或实时监测目标的位置。科学家能够用它来勘测和监视珊瑚礁。海上潜水员能够利用湿PC来寻找矿藏，开采其他未被发现的宝藏。

湿PC还可以被应用在教育、开发和旅游等方面。潜水员能够利用湿PC在海底礁石间领航，为水下的游客提供数字指路。

可佩戴式计算机所面临的挑战

可佩戴式计算机的创造是具有挑战性的，主要问题是如何生产出不太昂贵的产品。大多数器件对于普通用户来说太贵了。在开发方面，设计者和工程人员正在继续努力提高其功能性和舒适性。

“对于我们而言，一个巨大的挑战是在保证实用性、功能不断增强和体积不断减小的同时，继续使可佩戴式计算机的用户界面更友好。”Logic董事长Danny Cunagin说。

当前大多数可佩戴式计算机都使用标准的台式机的规格——奔腾处理器、Windows操作系统，创建了一个简单、细致而且舒服的用户界面。

Fast Reading Two Linux

Many users get annoyed by everyday troubles that still affect Windows systems. Whether it is a failing program, so why not use Linux instead?

Linux is a freely available and distributable look-alike of UNIX developed primarily by Linus Torvalds(Pic 2.2) at the University of Helsinki in Finland. Linux was further developed with the help of many UNIX programmers and wizards across the Internet.

Pic 2.2 Linus Torvalds

Linux is small, fast, and flexible. Linux has been publicly available since about November 1991. Version 0.10 went out at that time, and version 0.11 followed in December 1991. There are very few bugs now, and in its current state, Linux is most useful for people who are willing to port and write new codes.

Fast Reading Two

Linux is in a constant state of development. Linux is cheaper to get than most commercially available UNIX systems. If you have the patience and access to the Internet, the only price you pay for Linux is your time. For a nominal fee anywhere from ＄30 to ＄200, you can save yourself some time and get CD-ROM or floppy-disk distributions from several commercial vendors. The most important advantage of using Linux is that you get to work with a real kernel. All the kernel source code is available for Linux, and you will have the ability to modify it to suit your needs. Looking at the kernel code is an educational experience in itself.

There are different options in the Linux space. Red Hat and SuSe are two of the most popular ones. Red Hat Linux is a popular version of Linux that comes with the GNOME desktop environment. GNOME displays pictures on the screen to help you perform tasks. After that, selecting the software you want to use will also consume some time, but you can be sure to find something that fits your needs.

Unlike Windows, most Linux distributions have a very modular structure, which means that you can choose to install a very slim basic system and a few, specialized modules only. As a result, a minimum installation does not require more than a few hundred megabytes.

Linux is a free UNIX clone that supports a wide range of software such as Xwindow systems, GNU, C/C++ Compiler and TCP/IP. It's a versatile, very UNIX—like implementation of UNIX, freely distributed by the terms of the GNU' General Public License. Linux is also very closely compliant with the OPOSIX.1 standard, so porting applications between Linux and UNIX systems is a snap.

Linux is a clone of the UNIX operating system that runs on Intel 80x86-based machines, where x is 3 or higher. Linux is also very portable and flexible because it has now been ported to DEC Alpha, PowerPC, and even Macintosh machines. And progress is being made daily by Linux enthusiasts all over the world to make this free operating system available to all the popular computing machines in use today. Because the source code for the entire Linux operating system is freely available, developers can actually spend time porting the code, regardless of the payment of licensing fees.

Documentation for the many parts of Linux is not very far away, yet. The Linux Documentation Project (LDP) is an effort put together by many dedicated and very smart individuals to provide up-to-date, technically valuable information. All of this LDP information can be found on the Internet at various Linux source repositories. Each "HOWTO" document for Linux is the result of effort from many Linux enthusiasts. The original authors of these documents are usually also the core Linux developers who have put in hours of time and effort while struggling with new features of Linux. These individuals are the ones who deserve the credit.

.End.

参考译文 | 快速阅读2 Linux

许多用户为Windows操作系统的日常问题而感到烦恼。也许这是一个有缺陷的程序，为什么我们不用Linux来代替它呢？

Linux是一个酷似UNIX而且可以免费获取的操作系统，最初是由芬兰赫尔辛基大学的Linus Torvalds（图2.2）研制的。后来Linux通过因特网，在许多UNIX程序设计人员和精英的帮助下得到了进一步开发。

Linux具有小、快、灵活的特点。大约在1991年11月，Linux已开始公开发行。那时0.10版被推出，接着在当年的12月份0.11版被推出。当前版本的Linux错误极少，对于那些想移植和编写新代码的人来说，是非常有用的。

Linux处于不断开发之中。它比大多数商业化的UNIX系统便宜。如果你有耐心并可以访问因特网，只要付出一定的“时间”便可以购买到Linux。在任何地方，正常只需支付30美元至200美元，你便可以节约时间从一些商业售货机构获得CD-ROM，或是一些软盘的发行版。使用Linux的最大好处就是可以用一个真正的内核工作。所有的内核源代码Linux都可使用，并能根据用户的需求加以修改。研究内核代码本身就是自学的好方法。

Linux有不同的版本可供选择。Red Hat和SuSe是两种最流行的版本。Red Hat Linux是Linux的流行版本，它拥有GNOME桌面环境。GNOME在屏幕上显示图形来帮助你执行任务。然后，选择你希望用的软件将要耗费一些时间，但是你一定会找到满足你需求的产品。

与Windows不同，大部分的Linux发行版具有非常模块化的结构，这意味着你只需选择安装非常小的基本系统和几个专门模块。其结果是，最基础的安装所需空间不过几百兆。

Linux是UNIX的免费克隆版，它支持的软件十分广泛，如Xwindow系统、GNU、C/C++编译器和TCP/IP等。它是通用的，实现起来非常像UNIX，并根据GNU的公用许可证（GPL）协议免费发行。Linux非常接近OPOSIX.1标准，所以在Linux与UNIX间移植应用程序非常快捷。

Linux是UNIX操作系统的克隆版，它运行在基于Intel 80x86处理器的机器上，这里的x大于或等于3。Linux非常易于移植，灵活性很强，因为它现在已经被移植到如DEC Alpha、PowerPC，甚至是Mac机上了。全世界的Linux爱好者每天都在改进它，使这个免费的操作系统能运行在当今所有流行的机器上。因为全部的Linux操作系统的源代码都是免费的，所以开发者实际上只需花费时间来移植代码，不用考虑支付受益许可费的问题。

Linux许多部分的文档也还远未流传开。Linux的文档计划（LDP）是把众多富有奉献精神和极具智慧的个人的努力集中起来，以提供最新的技术和最有价值的信息的一项计划。所有的LDP文档信息都能在因特网上的各种Linux储存库中被找到。Linux的每个“HOWTO”文档都是Linux爱好者共同努力的结晶。这些文档的原作者，通常也是花费了许多时间和精力的核心Linux的开发者。他们为Linux的新特性而奋斗着，这些人理应享有赞誉。

Exercises

Ex 1 What is an operating system? Try to give a brief summary according to the passage.

Ex 2 **Fill in the table below by matching the corresponding Chinese or English equivalents.**

Kernel	
	应用程序接口
interface	
	操作系统
application software	
	平台

Ex 3 **Choose the best answer for each of the following statements according to the text we've learnt.**

1. An OS consists of one or more ________ that manage the operations of a computer.
   A. interfaces
   B. programs
   C. kernel techniques
   D. account numbers
2. ________ is estimated 88.9 percent total usage share on Web connected computers.
   A. Platform
   B. Media Player
   C. Graphical User Interface
   D. Microsoft Windows
3. The latest version of Microsoft is ________.
   A. Windows XP
   B. Windows 7
   C. Windows 10
   D. DOS
4. System software provides a ________ to other software.
   A. Platform
   B. Plug and Play
   C. Media Player
   D. Windows 95
5. Which of the following is not the example of operating systems?
   A. Windows
   B. Unix
   C. Linux
   D. 3ds Max

Part B Practical Learning

Task One

Task Two

Training Target

In this part, there are two tasks in English environment. You should complete these tasks in groups under the joint guidance of professional teachers and laboratory teachers, so as to train and improve yours ability to complete professional tasks in English environment.

Task One Download the Software (Windows 7 and Application Software) from the Internet

The first task is to download necessary software from the Internet. In this task, students must know the necessary software needed to finish the whole task.

There is some information about downloading software. The information can help students finish the task.

The needed software can be downloaded through web page: *www. baidu.com* or *www. google.com* (Pic 2.3, Pic 2.4).

Pic 2.3 baidu

Pic 2.4 google

You can click on the relevant options to enter into relevant pages for details (Pic 2.5, Pic 2.6).

Pic 2.5 Page for details (1)

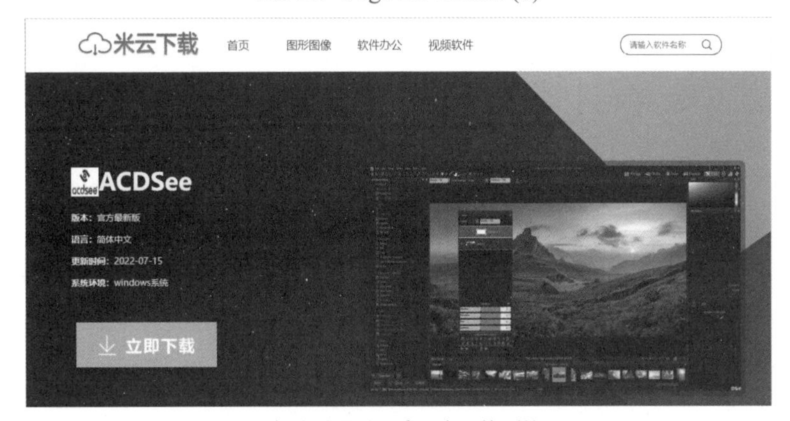

Pic 2.6 Page for details (2)

Task Two Install the Software

In this task, students must install the needed software which is downloaded from the Internet. Just run the setup program and follow the prompts (Pic 2.7).

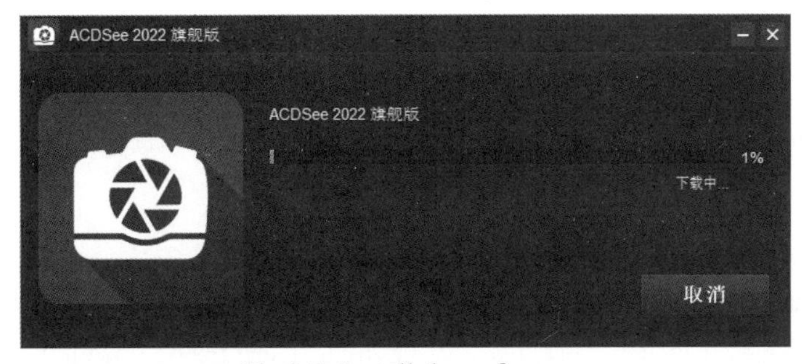

Pic 2.7 Install the software

Part C Occupation English

Occupation English

In this part, there is an English dialogue in real life and work environment. You will play the roles of A and B and read the dialogue aloud to practice your ability to use English.

Installing Operating System 安装操作系统

Character Setting: Computer Software Installer (A) and Customer (B)

角色设置：电脑软件安装人员（A）和客户（B）

A: Hello, can I help you?您好，请问有什么可以帮您吗？

B: I've just bought a new computer in your shop. I need to install the operating system and all kinds of application software.我刚在贵店买了一台新电脑，我需要安装一下操作系统和各类应用软件。

A: OK, would you need me to install them for you, or should I teach you how to install them?好的，您是需要我给您安装好了，还是我教您如何安装？

B: I would like you to guide me to install, so that I can install the software more conveniently in the future.我想让您指导我安装，这样将来我自己想要安装软件也方便一些。

A: OK, no problem.好的，没有问题。

B: That's great. Let's get started.那太好了，我们开始吧。

A: The latest Windows 10 system supports online upgrade, but the online upgrade method is slow and prone to unpredictable problems. We choose to install it with U disk. Here I need to explain: because we have made Windows 10 installation U disk, we can directly use U disk to install. If you don't have the Internet access to find out how to make the installation U disk, you can also contact us to guide you to make the installation U disk.最新Windows 10系统支持在线升级，但是在线升级的方式速度慢，而且容易出现不可预知的问题，我们选择用U盘安装。这里我需要说明一下：因为我们已经制作好了Windows 10安装U盘，所以我们可以直接用U盘安装。如果你无法上网查找如何制作安装U盘，也可以联系我们指导你制作安装U盘。

B: OK, I see.好的，我清楚了。

A: The installation steps are as follows: 安装步骤如下：

1. Insert the made Windows 10 installation U disk to the computer, and then restart the computer. Press the startup hotkeys F12, F11, Esc ceaselessly to select the boot from the U disk.在电脑上

插入制作好的Windows 10安装U盘，然后重启电脑。不停按F12、F11、Esc等启动热键选择从U盘启动。

2. Start the installation environment by booting from the USB flash disk.从U盘启动进入，开始启动安装环境。

3. Open the Windows setup window and click "Install Now" to start installing Windows 10.打开Windows安装窗口，然后单击“立即安装”开始安装Windows 10。

4. In the pop-up installation interface, click accept to install Windows 10 protocol, and then click next. You can follow the prompts step by step to install Windows 10 on the computer.在弹出的安装界面，点击接受安装Windows 10协议，然后再点击下一步，你可以根据提示一步步执行下去，就会把Windows 10完好地安装在电脑上了。

A: Do you know how to install the operating system now?你现在知道如何安装操作系统了吗？

B: Yes, after your demonstration, and I have searched the Internet for relevant information before. Now I understand how to install the operating system. I'll go back and install it myself. I'll consult you if I don't understand. Thank you very much!是的，经过您的演示，还有我之前也上网查找过相关资料，现在明白如何安装操作系统了。我回去自己安装一下，有不明白的地方我再咨询你们。非常感谢！

A: You are welcome! You are welcome to inquire at any time.不客气！欢迎您随时咨询。

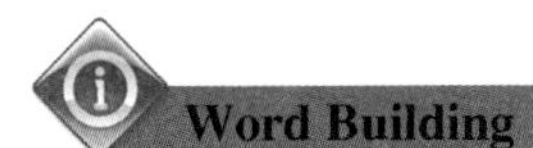

前缀/后缀由一个或几个字母组成，放在词根或单词之前/之后，组成一个新词。

(1) kilo-(前缀)：千

gram 克 ———— kilogram 千克

meter 米 ———— kilometer 千米

(2) sub-（前缀）：在……底下，子，次

head 标题 ———— subhead 小标题，副标题

directory 目录———— subdirectory 子目录

(3) en-（前缀)：使……

able 有才能的 ———— enable 使能够

close 关闭 ———— enclose 装入，附上

(4) -ment（后缀）：行为，状态，过程，手段及其结果

move 移动 ———— movement 运动

judge 判断 ———— judgement 判断

(5) -th（后缀）：动作，性质，过程，状态

true 真实的，真正的 ———— truth 事实，真理

weal 福利，幸福 ———— wealth 财富

Ex Translate the following words and try your best to guess the meaning of each word on the right according to the clues given on the left.

byte	字节（名词）	kilobyte ______
bit	位，比特（名词）	kilobit ______
way	路（名词）	subway ______
punish	惩罚（动词）	punishment ______
argue	争论，辩论（动词）	argument ______
large	大的（形容词）	enlarge ______
danger	危险（名词）	endanger ______
develop	发展（动词）	development ______
equip	装备（动词）	equipment ______
grow	生长（动词）	growth ______

Ex 1 What is software like? Try to give a brief summary of "*Application Software—OA*" in no more than five sentences.

Ex 2 What is application software? Try to give several examples.

Ex 3 What information do Smart Tags offer on the screen while you work?

Ex 4 Fill in the table below by matching the corresponding Chinese or English equivalents.

version	
	应用软件
wearable computer	
	文件
paste	
	网页
data processing	
	光盘只读存储器
XML	
	办公自动化

Ex 5 Choose the best answer for each of the following questions according to the text we've learnt.

1. Software is the set of ________ that tell a computer what to do.
   A. tools
   B. rules
   C. symbols
   D. instructions
2. Software can be categorized into two types: ________ and ________.
   A. bundled software, office software
   B. package software, custom software
   C. system software, application software
   D. freeware software, public-domain software
3. When you run your computer, ________ will be started first.
   A. operation system
   B. system software
   C. office software
   D. application software
4. When the user uses a new version of the Office to work, ________ offer(s) information and display(s) AutoCorrect, AutoFormat and Paste options.
   A. word processing system
   B. desktop publishing system
   C. Smart Tags
   D. voice recognition and synthesis
5. Three stages in the development of the OA system have accomplished the leap, which are ________.
   A. from data processing to information processing
   B. from information processing to knowledge processing
   C. from data processing to knowledge processing
   D. A and B

Project Three

LAN Setup and Connecting It to the Internet

Part A Theoretical Learning

Part B Practical Learning

Part C Occupation English

Part A Theoretical Learning

Training Target

In this part, our target is to improve the speed of reading professional articles and the comprehension ability of the reader. We have marked specialized key words and some flexible sentences. Try to grasp the main idea of each paragraph.

Skill One Foundation of Network

Computer networks are data **communication** systems made up of hardware and software. **A network is a collection of computers and devices connected together via communication devices such as cable telephone lines, modems, or other means. Sometimes a network is wireless; that is, it uses no physical lines or wires. When your computer is connected to computer network, users can share resources, such as hardware devices, software programs, data, and information.** Sharing resources saves time and money. For example, instead of purchasing one printer for every computer via a network, the network enables all of the computers to the access the same printer.

communication [kəˌmjuːnɪ'keɪʃn] *n.* 通信
via ['vaɪə] *prep.* 通过，经过
wireless ['waɪələs] *adj.* 无线的
physical ['fɪzɪkl] *adj.* 物理的

1. Types of Networks

Networks are often **classified** according to their **geographical** extent: LAN, MAN, WAN.

◇ **Local** Area Network (LAN)

A typical LAN spans a small area like a single building or a small campus and operates between 10 Mbps and 2 Gbps. Because LAN technologies cover short distances, they provide the highest speed connections among computers. **Ethernet** and **FDDI** are examples of standard LANs.

◇ **Metropolitan** Area Network (MAN)

MAN is a bigger version of a LAN in a city. It is smaller than a WAN but larger than a LAN. MAN is a public high-speed network, and runs at a speed of 100 Mbps or even faster, capable of voice and data transmission over a distance of up to 80 kilometers (50 miles).

◇ Wide Area Network (WAN)

WAN is sometimes called long haul network, providing communication over long distances. It can span more than one

classify ['klæsɪfaɪ] *v.* 分类
geographical [dʒɪə'græfɪkəl] *adj.* 地理的
local ['ləʊkl] *adj.* 地方的，当地的
Ethernet ['iːθənet] *n.* 以太网
FDDI 光纤分布式数据接口
metropolitan [ˌmetrə'pɒlɪtən] *adj.* 首都的，大城市的

Skill One

geographical area, often a country or continent. Usually WANs operate at slower speeds than LANs, and have much greater delay between connections. Typical speeds for a WAN range from 56Kbps to 155 Mbps. The Internet can be correctly regarded as the largest WAN in existence.

2. Topology

topology [tə'pɒlədʒɪ] *n.* 拓扑结构

A network architecture can be described in two ways: client-server and **peer-to-peer**. A client-server network is a network comprised of several workstations and one or more servers. In client-server networks, an administrator can control the **privileges** of each user. A peer-to-peer network is a type of network where all computers on the network have the **potential** to share resources that they have control over. All computers on the network can potentially act as both a client and a server. Because of this fact, there is no central control of the network and therefore this type of network structure is considered to be less secure and harder to manage than the client-server architecture.

peer-to-peer 对等网络

privilege ['prɪvəlɪdʒ] *n.* 特权

potential [pə'tenʃl] *n.* 潜能，潜力

Topology defines the structure of the network. There are five major topologies in use today: Bus, Ring, Star, Tree, and **Mesh**. Each is used for a specific network type, although some network types can use more than one topology.

mesh [meʃ] *n.* 网孔，网丝

◇ Bus: The simplest topology is the Bus (Pic 3.1). In the Bus, all the devices on the network are connected to a common cable. Normally, this cable is terminated at either end, and can never be allowed to form a closed loop.

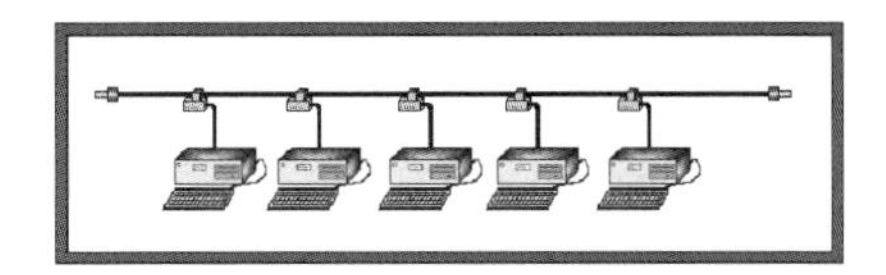

Pic 3.1 Bus topology

◇ Ring: A Ring topology (Pic 3.2) is very similar to a Bus. In a Ring, all the devices on the network are connected to a common cable which loops from machine to machine. After connecting the last machine on the network, the cable then returns to the first device to form a closed loop.

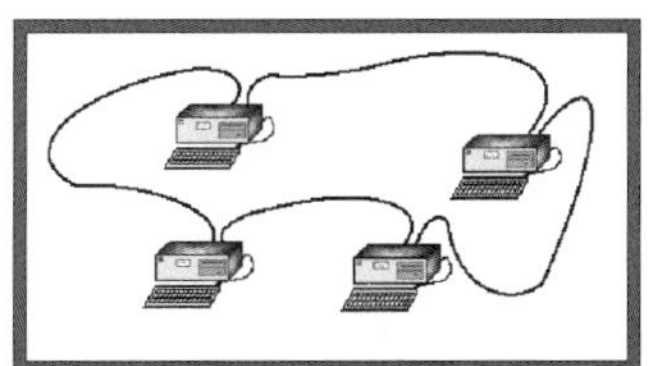

Pic 3.2 Ring topology

◇ Star: A Star topology (Pic 3.3) is completely different from either a Bus or a Ring. In a Star, each device has its own cable that connects the device to a common hub or concentrator. Only one device is permitted to use each port on the hub.

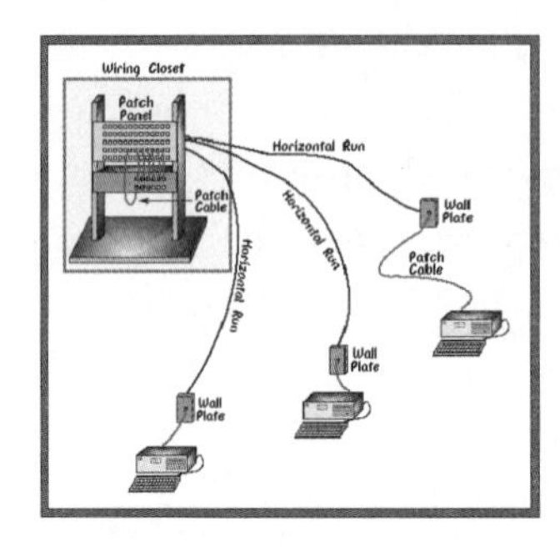

Pic 3.3 Star topology

◇ Tree: A Tree topology (Pic 3.4) can be thought of as a "Star of Stars" network. In a Tree network, each device is connected to its own port on a concentrator in the same manner as in a Star. However, concentrators are connected together in a **hierarchical** manner.

hierarchical [haɪə'rɑ:kɪkl] *adj.* 分等级的

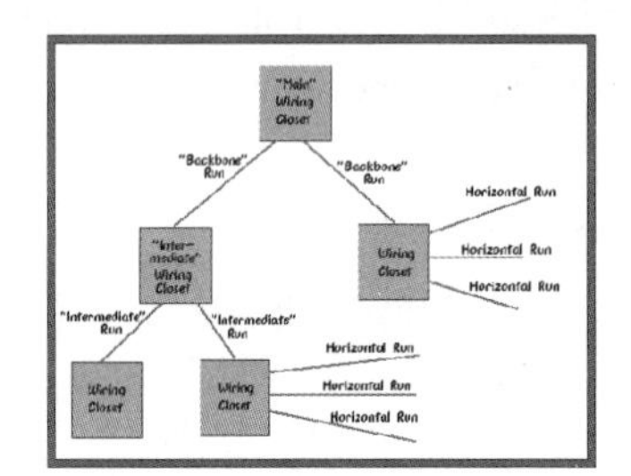

Pic 3.4 Tree topology

◇ Mesh: A Mesh topology consists of a network where every device on the network is physically connected to every other device on the network. This provides a great deal of performance and reliability. However, the complexity and difficulty of creating one increase **geometrically** as the number of nodes on the network increases. Pic 3.5 shows an example of a four-node Mesh network.

geometrical [dʒɪə'metrɪkl] *adj.* 几何的，几何学的

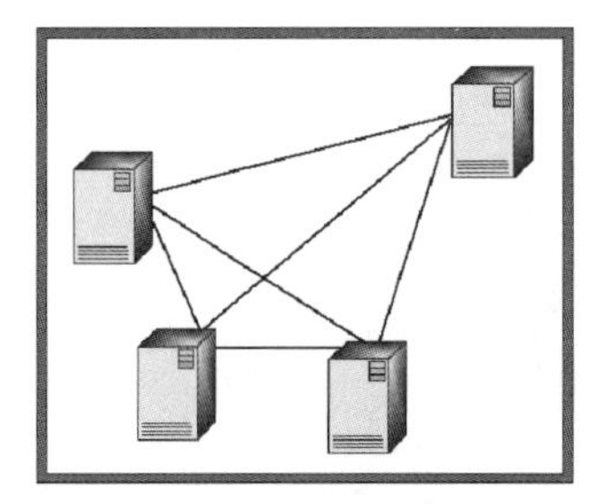

Pic 3.5 A four-node Mesh network

.End.

Key Words

communication *n.* 通信
wireless *adj.* 无线的
classify *v.* 分类
local *adj.* 地方的，当地的
FDDI 光纤分布式数据接口
topology *n.* 拓扑结构
privilege *n.* 特权
mesh *n.* 网孔，网丝
geometrical *adj.* 几何的，几何学的
via *prep.* 通过，经过
physical *adj.* 物理的
geographical *adj.* 地理的
Ethernet *n.* 以太网
metropolitan *adj.* 首都的，大城市的
peer-to-peer 对等网络
potential *n.* 潜能，潜力
hierarchical *adj.* 分等级的

参考译文 技能1 网络基础

计算机网络是由硬件和软件构成的数据通信系统。它通过通信设备，比如电话线、调制解调器或其他方式将一系列的计算机和设备连接起来。有时，网络也可以是无线的，也就是不需要物理线路或电缆。当你的计算机连接到网络上时，计算机用户就可以共享资源，比如硬件设备、软件程序、数据和信息。共享资源可以节省时间和金钱。举个例子吧，网络可以使处于同一网络内的计算机使用同一台打印机，而不用为网络中的每一台计算机都购买一台打印机。

1.网络类型

根据网络覆盖的地理范围，网络可以分为局域网、城域网和广域网。

◇ 局域网（LAN）

一个典型的局域网跨越的范围较小，诸如一幢大楼或者是一个小型校园，运行的速度在10 Mbps到2G bps之间。局域网技术覆盖较短的距离，从而提供了计算机间的最高速连接。以太网和FDDI就是典型的局域网。

◇ 城域网（MAN）

城域网是一种用于城市间的较大型局域网。它比广域网规模小，但比局域网规模大。城域网是一种公用的高速的计算机网络，运行速度为100 Mbps甚至更快，可以传输语音和数据，最大传输距离能达到80公里(50英里)。

◇ 广域网（WAN）

广域网有时被称为远程网，它能提供长距离的通信。广域网的跨度不止一个地区，经常包含一个国家或一个州。广域网的运行速度通常比局域网低，而且在连接之间有较大的传输延时。广域网的典型速度在56 Kbps到155 Mbps之间。因特网是现存最大的广域网。

2.拓扑结构

网络体系结构有两种类型：客户/服务器模式和对等网络模式。客户/服务器网络是一个包括几个工作站和一台或多台服务器的网络。在客户/服务器网络中，管理员能够控制每个用户的权限。在对等网络中，所有计算机都能共享它们控制的资源。网络中的计算机既是客户又是服务器。正是因为没有网络的集中控制，与客户/服务器模式相比，这种网络结构被认为缺少安全性，并且更难管理。

拓扑定义了网络的结构。目前主要有5种拓扑结构：总线型、环型、星型、树型和网型。一种拓扑结构适用于一个具体的网络，当然一些网络可使用不止一种拓扑结构。

◇ 总线型：最简单的拓扑结构就是总线型（图3.1）。在总线型网络中，所有的网络设备都连接到一条公用电缆上。这条电缆在两端终结，并且永远不会形成一个封闭的环路。

◇ 环型：环型拓扑（图3.2）与总线型拓扑非常相似。在环型网络中，所有的网络设备都连接到一条公共缆线上，缆线从一个设备连到另一个设备，再从最后一个设备连回到第一个设备，形成一个封闭的环路。

◇ 星型：星型拓扑（图3.3）完全不同于总线型或者环型拓扑。在星型网络里，每个设备都通过专用电缆连到集线器或中心控制器上。一台设备只允许连到集线器的一个端口上。

◇ 树型：树型拓扑（图3.4）可以被认为是一种“星型中的星型”网络。在树型网络内，每个设备就像在星型网络中那样连接到次级集线器中的一个端口上，再由次级集线器连到上级集线器上。

◇ 网型：在网型拓扑结构中，网络中的每个设备都与网络中其他所有设备有一条物理连接，这在很大程度上提高了网络的性能和可靠性。但是随着网络节点的增加，复杂性和安装的难度也几何级增大了。图3.5显示的是一个4个节点的网型网络。

Skill Two Network Devices

Skill Two

Network devices include all computers, media, interface cards and other equipment needed to perform data-processing and communications within the network. Let’s look at some typical network devices.

◇ **Network Interface Card (NIC)**

The network interface card provides the physical connection between the network and the computer workstation. Most NICs are internal, with the card fitting into an **expansion** slot inside the computer. NIC is a major factor in determining the speed and performance of a network. It is a good idea to use the fastest network card available for the type of workstation you are using. NICs are considered Layer 2 devices because each individual NIC throughout the world carries a **unique** code, called a Media Access Control (MAC) address. This address is used to control data communication for the host on the network.

expansion [ɪk'spænʃn] *n.* 扩展

unique [ju'ni:k] *adj.* 唯一的

◇ **Repeater**

One of the disadvantages of the type of cable that we primarily use, CAT5 UTP, is cable length. The maximum length for UTP cable in a network is 100 meters (approximately 328 feet). If you need to extend beyond the network limit, you must add a device to your network. This device is called a repeater. The repeater electrically **amplifies** the signal it receives and rebroadcasts it. Repeaters are networking devices that exit at Layer 1, the physical layer, of the OSI reference model.

repeater [ri'pɪ:tə(r)] *n.* 中继器

amplify ['æmplɪfaɪ] *v.* 放大，增强

◇ **Hub**

Generally speaking, hub is used when referring to the device that serves as the center of a network, as shown in Pic 3.6. The purpose of a hub is to **regenerate** and retime network signals. You will notice the characteristics of a hub are similar to the repeater's, which is why a hub is also known as the multi-port repeater. The difference is the number of cables that connect to the device. Where a repeater typically has only two ports, a hub generally has from four to twenty or more ports. Whereas a repeater receives a signal on one port and repeats it on the other, a hub receives a signal on one port and transmits it to all other ports.

regenerate [rɪ'dʒenəreɪt] *v.* 再生

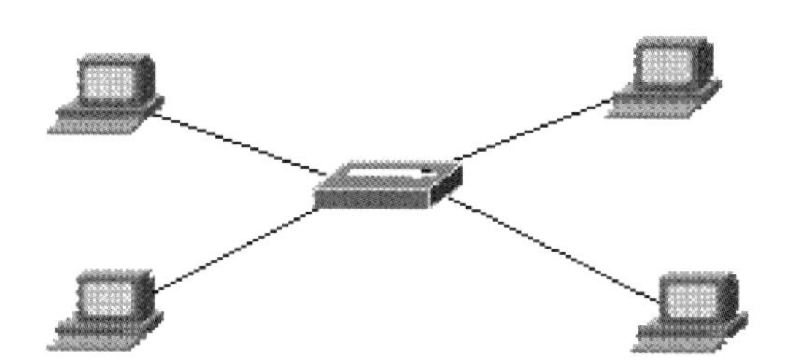

Pic 3.6 Hub as the center of a network

◇ **Bridge**

A bridge is a Layer 2 device that allows you to **segment** a large network into two smaller, more efficient networks. Bridges can be used to connect different types of cable, or physical topologies. They must, however, be used between networks with the same protocol. Bridges are **store-and-forward** devices. Every networking device has a unique MAC address on the NIC. Bridges **filter** networking traffic by only looking at the MAC address. Therefore, they can rapidly forward traffic representing any network layer protocol. Because bridges look only at MAC addresses, they are not concerned with network layer protocols. Consequently, bridges are concerned only with passing or not passing **frames**, based on their destination MAC addresses.

segment ['segmənt] *v.* 分割

store-and-forward 存储转发

filter ['fɪltə(r)] *v.* 过滤

frame [freɪm] *n.* 帧

◇ **Switch**

A switch is a Layer 2 device just as a bridge. In fact, a switch is sometimes called a multi-port bridge, just like a hub is called a multi-port repeater. Switches, at first glance, often look like hubs. Both hubs and switches have many connection ports. The difference between a hub and a switch is what happens inside the device. Switches make a LAN much more efficient. They do this by "switching" data only to the port to which the proper host is connected. In contrast, a hub sends the data to all its ports so that all the hosts have to see and process all the data. Pic 3.7 shows a switch and the symbol for a switch.

Pic 3.7 A switch and the symbol for a switch

◇ **Router**

The router is the first device you work with that is at the OSI network layer, also known as Layer 3. The router makes decisions based on network addresses as opposed to individual Layer 2 MAC addresses. The purpose of a router is to examine incoming packets, choose the best path for them through the network, and then switch them to the proper outgoing port. Routers are the most important traffic regulating devices on large networks. They enable virtually any type of computer to communicate with any other computer anywhere in the world! Pic 3.8 shows a router and the symbol for a router.

Pic 3.8 A router and the symbol for a router

◇ **Gateway**

A gateway can translate information between different network data formats or network architectures. It can translate TCP/IP to AppleTalk, so that computers supporting TCP/IP can communicate with Apple brand computers. Most gateways operate at the application layer.

◇ **Transmission Media**

Transmission media can be transferred in a wired or wireless method. The basic wired media are **twisted pair**, **coaxial** cable, and **optical fiber**. Wireless media have **terrestrial** microwave, satellite microwave and broadcast radio. Networking media are considered Layer 1 components of OSI model.

.End.

twisted pair 双绞线

coaxial [kəʊ'æksɪəl] *adj.* 同轴的

optical ['ɒptɪkl] *adj.* 光学的

fiber ['faɪbə(r)] *n.* 纤维

terrestrial [tə'restriəl] *adj.* 陆地的

Key Words

expansion *n.* 扩展
repeater *n.* 中继器
regenerate *v.* 再生
store-and-forward 存储转发
frame *n.* 帧
coaxial *adj.* 同轴的
fiber *n.* 纤维
unique *adj.* 唯一的
amplify *v.* 放大，增强
segment *v.* 分割
filter *v.* 过滤
twisted pair 双绞线
optical *adj.* 光学的
terrestrial *adj.* 陆地的

参考译文 技能2 网络设备

网络设备包括计算机、介质、网卡和网络中用于数据处理和通信的各种设备。下面就让我们看一些典型的网络设备吧！

◇ 网卡（NIC）

网卡为工作站和网络之间提供了一个物理连接。大多数网卡都是内置的，网卡一般插在计算机内的扩展槽中。网卡是影响网络速度和性能的一个主要因素，应尽可能为你的工作站配备最快速的网卡。网卡被认为是第二层的设备，因为世界上的每一块网卡都携带一个被称为媒体访问控制地址（MAC地址）的编码。这个地址被用来控制网上主机的数据通信。

◇ 中继器

我们主要使用的电缆（CAT5 UTP）的一个缺点是电缆长度有限制。在一个网络中UTP电缆的最大长度是100米（大约328英尺）。如果你想要延伸网络并超过这个界限，就必须在网络中添加一种设备。这种设备被称为中继器。中继器会将收到的信号进行放大然后再次传送。中继器是工作在OSI参考模型的第一层——物理层的网络设备。

◇ 集线器

一般来说，当我们需要一个设备作为网络中心时，就要利用集线器，如图3.6所示。集线器的作用是使网络信号再生并对其重新定时。你会注意到，集线器的特点与中继器类似，这就是集线器也被称为多端口中继器的原因。它们的区别是与设备相连接的电缆的数量不同。典型的中继器只有两个端口，而集线器的端口通常可有4到20个甚至更多。中继器从一个端口接收信号，重发到另一个端口上；而集线器从一个端口接收到信号后，却将它传送到其他所有的端口上。

◇ 网桥

网桥是第二层的设备，它允许你把一个大的网络分成两个更小、更高效的网络。网桥可以用来连接不同传输介质或不同拓扑结构的网络。不过，它们必须在使用相同协议的网络之间使用。网桥是一种存储转发设备。每一个联网设备的网卡上都具有唯一的MAC地址。网桥仅根据MAC地址来过滤网络流量。因此，它们能够迅速地将代表着任何网络层协议的流量向前传送。由于网桥只关注MAC地址，它们就不注重网络层的协议。因此，网桥只根据帧的目的MAC地址关注帧是否能够通过。

◇ 交换机

交换机像网桥一样也是第二层的设备。实际上，交换机经常被称为多端口网桥，就像集线器被称为多端口中继器一样。乍一看，交换机和集线器往往很像。无论是集线器还是交换机都有许多端口，它们的区别在设备的内部。交换机使LAN更有效。它们只将数据“交换”到相应主机的端口。相反地，集线器却将数据发送到它的全部端口上，这样所有的主机不得不看到并且处理所有的数据。图3.7显示的是一台交换机和它的表示符号。

◇ 路由器

路由器是工作在OSI第三层——网络层的第一个器件。路由器依据网络地址做出决策，而不是依据第二层的MAC地址。路由器的作用是检查流入的数据包，为它们选择通过网络的最佳路径，然后把它们交换到适当的输出端口。在大型网络中，路由器是最重要的流量控制设备。它们几乎能够使计算机与世界其他任何地方的计算机进行通信！图3.8显示的是一台路由器和它的表示符号。

◇ 网关

网关能在不同的网络数据形式或者网络体系结构之间传输信息。网关能把TCP/IP协议转换成AppleTalk协议，因此支持TCP/IP协议的计算机能与苹果牌计算机进行通信。大多数网关在应用层工作。

◇ 传输介质

传输介质可以是有线的，也可以是无线的。基本的有线传输介质是双绞线、同轴电缆和光纤。无线介质包括陆地微波、卫星微波和无线电波。网络介质被认为是OSI模型第一层的器件。

Fast Reading One

Fast Reading One TCP/IP Protocols

A wide variety of network protocol models exist, which are defined by many standard organizations worldwide and technology vendors over years of technology evolution and development. One of the most popular network protocol models is TCP/IP, which is the heart of Internet working communications.

The name TCP/IP refers to a suite of data communication protocols. The name is misleading because TCP and IP are only two of dozens of protocols that compose the suite. Its name comes from two of the more important protocols in the suite, the Transmission Control Protocol(TCP) and the Internet Protocol(IP). TCP/IP originated out of the investigative research into networking protocols that the Department of Defense (DoD) initiated in 1969. In the early 1980s, the TCP/IP protocols were developed. In 1983, they became standard protocols for ARPANET. Because of the history of the TCP/IP protocol suite, it is often referred to as the DoD protocol suite or the Internet protocol suite.

TCP/IP protocol suite includes more than 100 protocols. Now let's see some typical protocols.

◇ **Hypertext Transfer Protocol (HTTP)**

HTTP is an application-level protocol for distributed, collaborative, hypermedia information systems. It has been in use by the World Wide Web global information initiative since 1990. It is a stateless protocol which can be used for many tasks beyond its use for hypertext, such as name servers and distributed object management systems, through extension of its request methods, error codes and headers. A feature of HTTP is the typing and negotiation of data representation, allowing systems to be built independently of the data being transferred.

◇ **File Transfer Protocol(FTP)**

FTP enables file sharing between hosts. FTP uses TCP to create a virtual connection for control information and then creates a separate TCP connection for data transfers. The control connection uses an image of the TELNET protocol to exchange commands and messages between hosts.

◇ **Transmission Control Protocol (TCP)**

TCP supports the network at the transport layer. TCP specifies the format of the data and acknowledgement that two computers exchange to achieve a reliable transfer, as well as the procedure the computers use to ensure that the data arrive correctly. It specifies how TCP software distinguishes among multiple destinations on a given machine, and how communicating machines recover from errors like lost or duplicated packets. The protocol also specifies how two computers initiate a TCP stream transfer and how they agree when it is complete. Because TCP assumes little about the underlying communication system, TCP can be used with a variety of packet delivery systems, including the IP datagram delivery service. In fact, the large variety of delivery systems TCP can use is one of its strengths.

◇ **User Datagram Protocol (UDP)**

UDP supports the network at the transport layer. UDP provides the primary mechanism that application programs use to send datagrams to other application programs. UDP provides protocol ports used to distinguish among multiple programs executing on a single machine. UDP uses the underlying IP to transport a message from one machine to another, and provides the same unreliable, connectionless datagram delivery semantics as IP. It does not use acknowledgement to make sure messages arrive, it does not order incoming messages, and it does not provide feedback to control the rate at which information flows between the machines.

◇ **Internet Protocol (IP)**

IP provides support at the network layer. IP provides three important definitions. First, the IP protocol defines the basic unit of data transfer used throughout a TCP/IP Internet. Thus, it specifies the exact format of all data as it passes across a TCP/IP Internet. Second, IP software performs the routing function, choosing a path over which data will be sent. Third, IP includes a set of rules that embody the idea of unreliable packet delivery, meaning there is no guarantee that the data will reach the intended host. The datagram may be damaged upon arrival, out of order, or not arrive at all. IP is such a foundational part of the design that a TCP/IP Internet is sometimes called an IP-based technology.

◇ **Internet Control Message Protocol (ICMP)**

ICMP is used for network error reporting and generating messages that require attention. The errors reported by ICMP are generally related to datagram processing. ICMP only reports errors involving fragment 0 of any fragmented datagrams. The IP, UDP or TCP layer will usually take action based on ICMP messages. ICMP generally belongs to the IP layer of TCP/IP but relies on IP for support at the network layer. ICMP messages are encapsulated inside IP datagrams.

Evolution of TCP/IP technology is interwound with revolution of the global Internet. With millions of users at tens of thousands of sites around the world depending on the global Internet as part of their daily work environment, we have passed the early stage of development in which every user was also an expert, and entered a stage which few users understand the technology. Despite appearances, however, neither the Internet nor the TCP/IP protocol suite is static. Researchers are solving new networking problems, and engineers are improving the underlying mechanisms. In short, TCP/IP technology continues to evolve.

.End.

参考译文 快速阅读1 TCP/IP协议栈

许多国际标准化组织和技术开发商经过多年的技术开发和发展，形成了多种网络协议模型。最流行的网络协议之一就是TCP/IP协议，它是互联网通信的核心。

TCP/IP指的是一组数据通信协议。它有时被人们误解，因为TCP和IP只是协议栈诸多协议中的两个协议。这个名称来源于协议栈中两个比较重要的协议：传输控制协议(TCP)和网际协议(IP)。TCP/IP来源于1969年国防部（DoD）对网络协议的开发研究。20世纪80年代初期，TCP/IP得到了发展。1983年，它成为ARPANET的标准协议。由于TCP/IP协议栈的发展历史，它也经常被称作DoD协议或Internet协议栈。

TCP/IP协议栈包含了100多个协议。现在我们来看看其中的几个典型协议：

◇ **超文本传输协议**（HTTP）

HTTP是一个应用层的协议，用来发布、协调超文本信息。从1990年起它一直为万维网所使用。HTTP是一种通过请求方法、错误代码和报头的扩展格式来实现的无状态协议。它除了用于超文本传输之外，还用于很多其他的任务，例如，名称服务器和分布式的对象管理系统。HTTP的一个特点是允许系统独立地创建数据表示方法的类型和协商能力，而不依赖于所要传输的数据。

◇ **文件传输协议（FTP）**

FTP协议使在主机之间共享文件成为可能。FTP使用TCP为控制信息建立虚拟连接，然后为数据传输创建一个单独的TCP连接。控制连接使用TELNET协议在主机之间交换命令和消息。

◇ **传输控制协议（TCP）**

传输控制协议(TCP)工作在传输层。TCP指定了两台计算机之间为了进行可靠传输而交换的数据和确认信息的格式，还指定了计算机为了确保数据的正确到达而采取的步骤。该协议规定了TCP软件如何识别给定机器上的多个目标，及如何对类似分组丢失和分组重复这样的错误进行恢复。该协议还指出了如何在计算机之间实现发起TCP数据流的传输，以及完成后计算机如何同意开始传输数据。由于TCP对底层通信系统没有什么特殊要求，因此可用于包括IP数据包交付服务在内的多种数据包传输系统，这正是其强大功能的体现。

◇ **用户数据报协议（UDP）**

UDP工作在传输层。UDP提供应用程序之间传输数据报的基本机制。UDP提供的协议端口能够区分在一台机器上运行的多个程序。UDP使用底层的IP协议在各个机器之间传输报文，提供和IP一样不可靠、无连接的数据报传输服务。它没有使用确认机制来确保报文的准确到达，没有对传入的报文排序，也不提供反馈信息来控制机器之间信息传输的速度。

◇ **网际协议（IP）**

IP协议工作在网络层。IP协议提供了三个重要的定义。第一，IP协议定义了在整个TCP/IP互联网上数据传输所用的基本单元。因此，它规定了互联网上传输的数据的确切格式；第二，IP软件完成路由选择的功能，选择一个数据发送的路径；第三，IP包括了一组体现了不可靠数据传输的规则，这意味着不能保证数据一定能到达目的地。数据包可能被损坏，打乱顺序，或丢失。IP是TCP/IP互联网中最基本的部分，因此有时也称TCP/IP互联网为基于网际协议的技术。

◇ **网际控制报文协议（ICMP）**

网际控制报文协议(ICMP)用于网络错误报告及产生要求注意的消息。ICMP报告的错误通常与数据包处理有关。ICMP只报告与数据包第0分片有关的错误。IP、UDP或者TCP层通常都在ICMP报文的基础上工作。ICMP一般属于TCP/IP中的IP层，但是依赖网络层的IP协议。ICMP报文被封装在IP数据包里面。

TCP/IP技术的发展与整个互联网的发展密切相关。在整个互联网上有数以万计的站点，这些站点上有着数百万的用户，互联网已经成为这些用户日常工作环境的一部分。我们已经度过了那种每个用户同时也是专家的早期开发阶段，进入了一个新的阶段，即只有很少的用户明白其中的技术。然而，不管表面上怎样，实际上无论是互联网还是TCP/IP协议都不是固定不变的。研究人员不断解决新的网络问题，工程师们也在不断改进底层机制。总之，TCP/IP技术在不断发展。

Fast Reading Two　Windows Network Projector Overview

A Windows Network Projector is a display device, such as a conference room projector, that uses Remote Desktop Protocol over an IP network to display the desktop of a Windows Vista-based PC. Windows Embedded CE 6.0 comes with an OS design template that allows you to create Windows Network Projectors.

A Windows Network Projector built with CE 6.0 and later is focused on supporting business scenarios such as those in the following list.

* Microsoft PowerPoint presentations with simple animations and still image display.
* Displays to a single projector (one-to-one connection).
* Mirror or extended display.

The Network Projector utilizes Remote Desktop Protocol (RDP) for display capabilities. It can support wired or wireless network connections between the computer and the projector. It can support both ad hoc and infrastructure mode for wireless networks.

The Windows Embedded CE technologies behind Windows Network Projectors can allow you to build a number of different device types.

The following diagram (Pic 3.9) shows a direct implementation of Windows Network Projector built into a new or existing projector design. With this integrated support, the projector provides the capability of being discovered and connected to by a Windows Vista-based PC. This example shows the Windows Network Projector used with an infrastructure network connection.

Pic 3.10 shows the projector used with an ad hoc network connection.

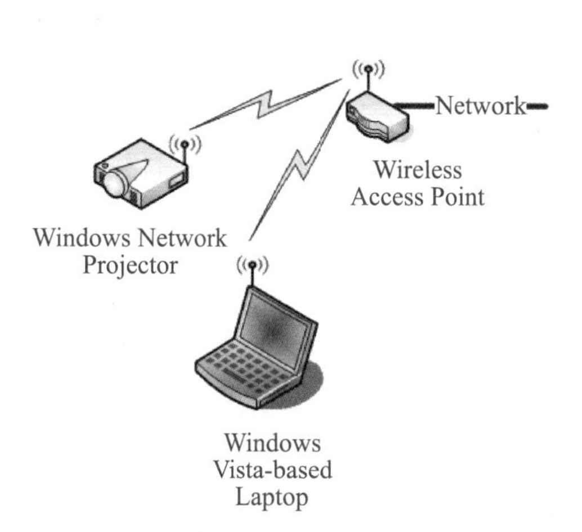

Pic 3.9 The projector used with an infrastructure network connection

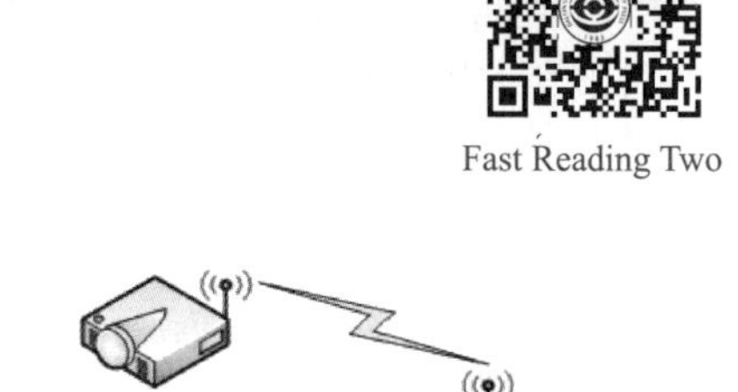

Pic 3.10 The projector used with an ad hoc network connection

参考译文 快速阅读2 Windows网络投影仪概述

Windows网络投影仪是一个显示设备,如会议室投影仪,使用IP网络中的远程桌面协议来显示装有Windows Vista系统的个人计算机的桌面。Windows Embedded CE 6.0内含操作系统平台,允许你创建Windows网络投影仪。

Windows网络投影仪内置CE 6.0,后来主要用于支持业务模式,具体内容如下:

*微软PowerPoint演示:有简单的动画,但仍是静态图像显示;

*单个投影仪展示(一对一的连接);

*镜像或扩展显示。

网络投影仪使用远程桌面协议(RDP)显示功能,支持由有线或无线网络连接的电脑和投影仪,并同时支持无线自组网和无线基础设施模式。

Windows网络投影仪背后的Windows Embedded CE技术可以允许你构建许多不同的设备类型。

图3.9直观地展现了Windows网络投影仪与新的或现有的投影仪连用的设计。在这种综合保障下,投影仪就可以被装有Windows Vista系统的电脑发现,并连接到电脑上。该例子说明的是Windows网络投影仪与基础网络结构连用的情况。

图3.10则显示了投影仪可以用于自组网连接。

Exercises

Ex 1 How many types of network are there? What are they? Can you give some examples about them?

Ex 2 Fill in the table below by matching the corresponding Chinese or English equivalents.

Ethernet	
	客户/服务器
FDDI	
	对等网络
topology	
	电缆
Mesh network	
	城域网
Ring network	

Ex 3 Choose the best answer for each of the following statements according to the text we've learnt.

1. Each computer or shared device found on the network is known as a ________.
   A. server
   B. workstation
   C. network node
   D. bridge
2. A ________ is a special computer on a network that provides and controls services (resources) for other computers on the network to use.
   A. workstation
   B. server
   C. bridge
   D. switch
3. ________ have limited access to the resources found on the network.
   A. Clients
   B. Administrators
   C. Nodes
   D. Servers
4. Every network has a "shape" which is normally referred to its ________.
   A. Bus
   B. topology
   C. Star
   D. Tree
5. In a ________ topology, each device has its own cable running to connect the device to a common hub or concentrator.
   A. Bus
   B. Mesh
   C. Star
   D. Tree

Part B Practical Learning

Task One

Task Two

Training Target

In this part, there are two tasks in English environment. You should complete these tasks in groups under the joint guidance of professional teachers and laboratory teachers, so as to train and improve yours ability to complete professional tasks in English environment.

Task One Connect the Computer to the Internet

In this task, students must connect their computers to the Internet through the campus network. Students can use cable to connect their computers to the laboratory's hub.

Task Two Construct LAN in English Environment

In this task, students must construct a Local Area Network in English environment through the campus network.

There is some information about the construction of a LAN. The information will help the students accomplish the task.

To build a Local Area Network, you need to set both of the computers to have an IP address in the same subnet. For example, you can set one of your computers IP addresses to be 192.168.0.1 and the other as 192.168.0.2.

Make sure you use the subnet mask 255.255.255.0. By using this subnet mask, you can use any address from 192.168.67.1 to 192.168.67.254.

You can look at the example (Pic 3.11) for a visualization to see how the TCP/IP settings should be set up.

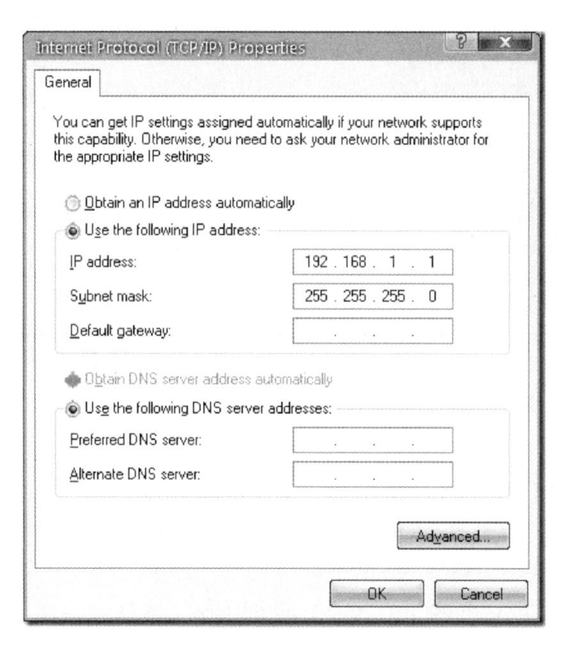

Pic 3.11 Internet Protocal (TCP/IP) Properies (1)

The second computer should be set up like Pic 3.12:

Pic 3.12 Internet Protocal (TCP/IP) Properies (2)

By keeping both IP addresses in the same subnet, they can "talk" to each other. In the example above, you can use 192.168.1.1 through to 192.168.1.254 if you keep the subnet mask the same at 255.255.255.0.

Once this is set up, you can share resources on both machines and be able to access them by launching the run window and entering the other computer's IP address in this format \\192.168.1.1, like Pic 3.13:

Now, if you get a window that looks like Pic 3.14 instead of Pic 3.13, you need to set up your folder's security. You can do this by right-clicking on the folder you're trying to share or access and choosing Sharing and Security. From here, you can give the appropriate permissions.

Pic 3.13 The run window

Pic 3.14 set up the folder's security

Part C Occupation English

Occupation English

Training Target

In this part, there is an English dialogue in real life and work environment. You will play the roles of A and B and read the dialogue aloud to practice your ability to use English.

How to Make Effective Use of E-learning 怎样利用网络有效学习

Character Setting: College Student （A）, College Student （B）

角色设置：大学生（A），大学生（B）

A: Hi, Li Ming. I saw you went to bed late yesterday. What were you doing online? Did you stay up late and play games? That's bad for your health!嗨，李明，我看你昨天睡得很晚，你在网上干什么？难道你是熬夜打游戏？那对身体很不好啊！

B: Hi, Wu Dong, I didn't play games. I have wanted to take National Computer Rank Examination (level Ⅱ). Last night I was on the Internet to find information and watch video learning, doing a bit late.嗨，吴东，我没打游戏，我要考国家计算机等级考试（二级）。昨天晚上我在网上查找资料，看视频学习，看得有点晚了。

A: I admire you. It's a matter of business to take National Computer Rank Examination(level Ⅱ). It's helpful to get a few certificates before graduation.我佩服你，考国家计算机等级考试（二级）是个正事。毕业前拿几个证书，对以后找工作还是有帮助的。

B: Of course. Are you interested in learning together?那当然了，你有没有兴趣一起学呀？

A: Thank you. I'm preparing for the exam of Internet of things certifier. I've heard that there is a shortage of talents in this field. The employment is relatively good, and the salary is also very high. 谢谢，我正准备物联网认证师的考试，听说这方面人才很缺，就业比较好，且工资也很高。

B: Oh, I didn't expect that, wow, you have great foresight! Indeed, the future of the Internet of things is very broad, much broader than the National Computer Rank Examination(level Ⅱ). 啊，没想到哇，你很有远见啊！的确，物联网方面以后的前景很广，比这个国家计算机等级考试（二级）广多了。

A: No, it's just a different direction for us to prepare.不，咱们只是准备的方向不同而已。

B: How do you learn?你用什么方式学习呢？

A: Online, of course! Online resources are rich. There're various forms, audio, videos, texts, pictures and other materials, as well as real questions over the years, other people's review and examination experience. Wow, it's countless, and also very convenient!当然是在线了！网上资源丰富，形式多样，有音频、视频、文字、图片等多种资料，还有历年真题、别人的复习和考试心得。哇，数不胜数，而且还很方便！

B: So am I. Let's work together to make good use of network resources and serve our own learning. Wish us all good results!我也一样。让我们一起努力，利用好网络资源，为自己的学习服务。预祝我们都取得好成绩！

A: I wish us all good results！预祝我们都取得好成绩！

Word Building

前缀/后缀由一个或几个字母组成，放在词根或单词之前/之后，组成一个新词。

(1) -ure（后缀）：行为或行为的结果、状态

fail 失败 ———— failure 失败

press 压、按 ———— pressure 压力

(2) -or（后缀）： ……的人

visit 参观，访问 ———— visitor 来访者

invent 发明 ———— inventor 发明者

(3) -ous（后缀）：有……的

danger 危险 ———— dangerous 危险的

fame 名声 ———— famous 著名的

(4) -ful(后缀)：充满……的，具有……性质的

use 用处 ———— useful 有用的

Ex Try your best to guess the meaning of each word on the right according to the clues given on the left.

expand	扩展（动词）	expansion ________
geometry	几何学（名词）	geometrical ________
repeat	重复（动词）	repeater ________
route	路（名词）	router ________
power	能力（名词）	powerful ________
flex	弯曲（名词）	flexure ________
essence	本质、精华（名词）	essential ________
magnetism	磁力（名词）	electromagnetism ________

Exercises

Ex 1 How many network devices do you know? Can you summarize the characteristics of them?

Ex 2 Fill in the table below by matching the corresponding Chinese or English equivalents.

twisted pair	
	交换机
repeater	
	路由器

optical fiber	
	集线器
OSI	
	网卡
UTP	
	物理地址

Ex 3 Choose the best answer for each of the following statements according to the text we've learnt.

1. NICs are considered Layer 2 devices because each individual NIC throughout the world carries a unique code, called _________.
   A. binary impulses
   B. serial format
   C. code of network interface card
   D. Media Access Control Address
2. One of the disadvantages of the type of cable that we primarily use, CAT 5 UTP , is _________.
   A. the cost of this kind of medium
   B. difficult to install
   C. cable length
   D. too complicated
3. Consequently, bridges are concerned only with passing or not passing frames, based on _________.
   A. Internet Protocol address
   B. logical address
   C. their destination MAC address
   D. the format of the frame
4. The purpose of a router is to examine incoming packets, choose the best path for them through the network, and then _________.
   A. let them wait there for some time
   B. change the information coming from the application layer
   C. switch them to the proper outgoing ports
   D. put them to another router
5. _________ is the best transfer media at present.
   A. Optical fiber　　B. Coaxial cable
   C. Twisted pair　　D. Radio
6. _________ can select the best path to route a message.
   A. NIC　　B. Bridge
   C. Switch　　D. Router

Project Four

Information Query and Analysis

Part A Theoretical Learning

Part B Practical Learning

Part C Occupation English

Part A Theoretical Learning

Training Target

In this part, our target is to improve the speed of reading professional articles and the comprehension ability of the reader. We have marked specialized key words and some flexible sentences. Try to grasp the main idea of each paragraph.

Skill One Big Data

Big data is a field that treats ways to analyze, systematically extract information from, or otherwise deal with data sets that are too large or complex to be dealt with by traditional data-processing application software. Data with many cases (**rows**) offer greater statistical power, while data with higher complexity (more **attributes** or **columns**) may lead to a higher false discovery rate. Big data challenges include capturing data, data storage, data analysis, searching, sharing, **transfering**, **visualization**, querying, updating, information **privacy** and data source. Big data was originally **associated** with three key concepts: **volume, variety**, and **velocity**. When we handle big data, we may not sample but simply observe and track what happens. Therefore, big data often includes data with sizes that **exceed** the capacity of traditional software to process within an acceptable time and value.

Current usage of the term big data tends to refer to the use of predictive analytics, user behavior analytics, or certain other advanced data analytics methods that extract value from data, and seldom to a particular size of data set. "There is little doubt that the quantities of data now available are indeed large, but that's not the most relevant characteristic of this new data ecosystem." Analysis of data sets can find new correlations to "spot business trends, prevent diseases, combat crime and so on." Scientists, business executives, practitioners of medicine, advertising and governments alike regularly meet difficulties with large data-sets in areas including Internet searches, **Fintech**, **urban** informatics, and business informatics. Scientists encounter limitations in e-Science work, including **meteorology**, genomics, connectomics, complex physics **simulations**,

big data 大数据
systematically [ˌsɪstə'mætɪkli] *adv.* 有系统地；有组织地
extract [ɪk'strækt] *v.* 取出
row [rəʊ] *n.* 行
attribute [ə'trɪbjuːt] *n.* 属性
column ['kɒləm] *n.* 列
transfer [træns'fɜː(r)] *v.* 转移
visualization [ˌvɪʒʊəlaɪ'zeɪʃn] *n.* 形象化，可视化
privacy ['prɪvəsi] *n.* 隐私
associate [ə'səʊsieɪt] *v.* 关联
volume ['vɒljuːm] *n.* 数据规模，卷；册；体积；容积
variety [və'raɪəti] *n.* 数据类型的多样式
velocity [və'lɒsəti] *n.* 速度
exceed [ɪk'siːd] *v.* 超过
Fintech 金融科技
urban ['ɜːbən] *adj.* 城市的
meteorology [ˌmiːtiə'rɒlədʒi] *n.* 气象学；气象
simulation [ˌsɪmju'leɪʃn] *n.* 模拟；仿真

biology and environmental research.

Definition

The term has been in use since the 1990s, with some giving credit to John Mashey for popularizing the term. Big data usually includes data sets with sizes beyond the ability of commonly used software tools to capture, curate, manage, and process data within a tolerable elapsed time. Big data **philosophy encompasses** unstructured, **semi-structured** and structured data, however the main focus is on unstructured data. Big data "size" is a constantly moving target, as of 2012 ranging from a few dozen terabytes to many zettabytes of data. Big data requires a set of techniques and technologies with new forms of integration to reveal insights from data-sets that are diverse, complex, and of a massive scale.

philosophy [fə'lɒsəfi] *n.* 哲学

encompass [ɪn'kʌmpəs] *vt.* 包围；包含；完成

semi-structured [ˌsemi'strʌktʃəd] *adj.* 半结构化的

Characteristics

Big data can be described by the following characteristics:

Volume

The quantity of generated and stored data. The size of the data determines the value and potential **insight**, and whether it can be considered big data.

insight ['ɪnsaɪt] *n.* 洞察力

Variety

Skill One

The type and nature of the data. The earlier technologies like RDBMSs were capable to handle structured data efficiently and effectively. However, the change in type and nature from structured to semi-structured or unstructured challenged the existing tools and technologies. The Big Data technologies evolved with the prime intention to capture, store, and process the semi-structured and unstructured (variety) generated with high speed(velocity), and huge in size (volume). Later, these tools and technologies were also utilized for structured data but preferable for storage. The processing of structured data was still kept as optional, either using big data or traditional RDBMSs. This helps people who analyze it effectively use the resulting insight from the data collected from social media, log files, and sensors, etc. Big data draws from text, images, audio, video; plus it completes missing pieces through data **fusion**.

fusion ['fju:ʒn] *n.* 联合

Velocity

The speed at which the data are generated and processed to meet the demands and challenges that lie in the path of growth and development. Big data is often available in real-time. Compared to small data, big data is produced more continually. Two kinds of velocity related to big data are the frequency of generation and the frequency of handling, recording, and publishing.

Veracity

It is the extended definition for big data, which refers to the data quality and the data value. The data quality of captured data can vary greatly, affecting the accurate analysis.

Other important characteristics of Big Data are:

Exhaustive

Whether the entire system (i.e.,=all) is captured or recorded.

Fine-grained and uniquely lexical respectively, the proportion of specific data of each element per element collected and if the element and its characteristics are properly indexed or identified.

Relational

If the data collected contains common fields that would enable a conjoining, or meta-analysis, of different data sets.

Extensional

If new fields in each element of the data collected can be added or changed easily.

Scalability

If the size of the data can expand rapidly.

Value

The utility that can be extracted from the data.

Variability

It refers to data whose value or other characteristics are shifting in relation to the context in which they are being generated.

.End.

exhaustive [ɪg'zɔ:stɪv] *adj.* 全面彻底的；详尽无遗的

relational [rɪ'leɪʃənl] *adj.* 有关的

extensional [ɪk'stenʃənl] *adj.* 外延的

scalability [ˌskeɪlə'bɪləti] *n.* 可量测性

value ['vælju:] *n.* 重要性

variability [ˌveəriə'bɪləti] *n.* 变化性

Key Words

Big data 大数据
extract *v.* 取出
attribute *n.* 属性
transfer *v.* 转移
privacy *n.* 隐私
volume *n.* 数据规模，卷；册；体积；容积
velocity *n.* 速度
Fintech 金融科技
meteorology *n.* 气象学；气象
philosophy *n.* 哲学
semi-structured *adj.* 半结构化的
systematically *adv.* 有系统地，有组织地
row *n.* 行
column *n.* 列
visualization *n.* 形象化，可视化
associate *v.* 关联
variety *n.* 数据类型的多样式
exceed *v.* 超过
urban *adj.* 城市的
simulation *n.* 模拟；仿真
encompass *vt.* 包围；包含；完成
insight *n.* 洞察力

Key Words

fusion *n.* 联合
relational *adj.* 有关的
scalability *n.* 可测量的
variability *n.* 变化性
exhaustive *adj.* 全面彻底的；详尽无遗的
extensional *adj.* 外延的
value *n.* 重要性

参考译文 技能1 大数据

大数据是一个研究如何分析、系统地提取信息或以其他方式处理传统数据处理应用软件无法处理的庞大或复杂数据集的领域。海量的（有较多阵列）数据能提供更大的统计能力，而较复杂（多属性或阵列）的数据也许会导致更高的错误发现率。大数据挑战包括捕获数据，数据存储，数据分析，搜索，共享，传输，可视化，查询，更新，信息隐私和数据源。大数据起初与三个关键概念有关：海量的数据规模，多样的数据类型和快速的数据流转。当我们处理大数据时，我们可能不会采样，而只是观察并跟踪会发生什么。因此，大数据通常包含大小超出传统软件在可接受的时间和价值内处理的数据。

当前，大数据一词的使用往往是指使用预测分析，用户行为分析或某些其他高级数据分析方法来从数据中提取价值，而很少使用特定大小的数据集。“毫无疑问，现在可用的数据量确实很大，但这并不是这个新数据生态系统最相关的特征。”数据集的分析结果能够在“洞察商业趋势、疾病预防及打击犯罪等方面”找到新的关联。科学家、企业管理者，医务、广告及政府从业者等都常常会在大数据集领域遇到困难，例如互联网搜索、金融科技、城市信息学及商业信息学。科学家们遇到的局限性的e-Science工作，包括气象、基因组学、连接组学、复杂的物理模拟、生物学和环境研究。

定义

该术语自20世纪90年代开始使用，有人称赞John Mashey推广了该术语。大数据通常包含的数据集的大小超出了常用软件工具在可容忍的经过时间内捕获、整理、管理和处理数据的能力。大数据哲学包含非结构化、半结构化和结构化数据，而主要关注于非结构化数据。大数据“大小”是一个不断移动的目标，如2012的范围从几十兆字节到许多字节的数据。大数据需要一套具有新的集成形式的技术和科技，以揭示来自多样、复杂和大规模的数据集的见解。

特征

大数据可以用以下特征来描述：

数据规模

生成和存储的数据量。数据的大小决定了价值和潜在的洞察力，以及它是否可以被认为是大数据。

数据类型的多样性

数据的类型和性质。像关系数据库管理系统这样的早期技术能够有效地处理结构化数据。然而，从结构化到半结构化或非结构化的类型和性质的变化对现有的工具和技术提出了挑战。大数据技术的发展主要是为了捕获、存储和处理以高速（速度）和大规模（容量）生成的半结构化和非结构化(种类)。后来，这些工具和技术也被用于结构化数据，但更适用于存储。结构化数据的处理仍然是可选的，要么使用大数据，要么使用传统的关系数据库管理系统。这有助于分析人员有效地利用从社交媒体、日志文件和传感器等收集的数据中获得的洞察力。大数据来源于文本、图像、音频、视频；此外，它通过数据融合完成缺失的部分。

数据流转速度

数据生成和处理的速度，以满足增长和发展道路上的需求和挑战。大数据通常是实时可用的。与小数据相比，大数据的产生更加连续。与大数据相关的两种速度是生成的频率以及处理、记录和发布的频率。

准确性

它是对大数据的扩展定义，指的是数据质量和数据价值。捕获数据的数据质量可能会有很大差异，影响准确的分析。

大数据的其他重要特征包括：

详尽的

整个系统是否被捕获或记录。

细粒度和唯一词法分别是每个元素的特定数据在收集的每个元素中所占的比例，以及该元素及其特征是否被正确索引或识别。

有关系的

是否收集到的数据包含公共字段，这些字段可以实现不同数据集的联合或元分析。

外延的

是否可以轻松添加或更改所收集数据的每个元素中的新字段。

可扩展性

数据的规模是否能够迅速扩大。

价值

可以从数据中提取的实用程序。

变化性

它指的是其值或其他特征相对于生成它们的上下文发生变化的数据。

Skill Two Cloud Computing

Cloud computing is the on-demand availability of computer system resources, especially data storage (cloud storage) and computing power, without direct active management by the user. The term is generally used to describe data centers available to many users over the Internet. Large clouds, **predominant** today, often have functions distributed over multiple locations from central servers. If the connection to the user is relatively close, it may be designated an edge server.

Clouds may be limited to a single organization (**enterprise clouds**), or be available to many organizations (**public clouds**).

cloud computing 云计算

predominant [prɪ'dɒmɪnənt] *adj.* 占优势的

enterprise cloud 企业云

public cloud 公共云

Cloud computing relies on sharing of resources to achieve coherence and economies of scale.

History

Cloud computing was **popularized** with Amazon.com releasing its Elastic Compute Cloud product in 2006.

popularize ['pɒpjələraɪz] *vt.* 普及

References to the phrase "cloud computing" appeared as early as 1996, with the first known in a Compaq internal document.

The cloud symbol was used to represent networks of computing equipment in the original ARPANET by as early as 1977, and the CSNET by 1981—both **predecessors** to the Internet itself. The word cloud was used as a metaphor for the Internet and a standardized cloud-like shape was used to **denote** a network on telephony **schematics**. With this **simplification**, the **implication** is that the specifics of how the **endpoints** of a network are connected are not relevant to understanding the **diagram**.

predecessor ['pri:dɪsesə(r)] *n.* 前辈；前任；前身
denote [dɪ'nəʊt] *v.* 表示
schematic [ski:'mætɪk] *n.* 图表
simplification [ˌsɪmplɪfɪ'keɪʃn] *n.* 简化
implication [ˌɪmplɪkeɪʃn] *n.* 暗示
endpoint ['endˌpɔɪnt] *n.* 端点
diagram ['daɪəgram] *n.* 图解

The term cloud was used to refer to platforms for distributed computing as early as 1993, when Apple spin-off General Magic and AT&T used it in describing their (paired) Telescript and PersonaLink technologies. In Wired's April 1994 feature "Bill and Andy's Excellent Adventure Ⅱ", Andy Hertzfeld commented on Telescript, General Magic's distributed programming language:

"The beauty of Telescript is that now, instead of just having a device to program, we have the entire Cloud out there, where a single program can go and travel to many different sources of information and create a sort of a virtual service. No one had conceived that before. The example Jim White the designer of Telescript, X.400 and ASN. uses now is a date-arranging service where a software agent goes to the flower store and orders flowers and then goes to the ticket shop and gets the tickets for the show, and everything is communicated to both parties."

Skill Two

Deployment models

Private cloud

Private cloud is cloud **infrastructure** operated solely for a single organization, whether managed internally or by a third party, and hosted either internally or externally. **Undertaking** a private cloud project requires **significant engagement** to **virtualize** the business environment, and requires the organization to reevaluate decisions about existing resources. It can improve business, but every step in the project raises security issues that must be addressed to prevent serious **vulnerabilities**. Self-run data centers are generally capital intensive.

private cloud 私有云
infrastructure ['ɪnfrəstrʌktʃə(r)] *n.* 基础结构
undertake [ˌʌndə'teɪk] *v.* 保证；任务
significant [sɪg'nɪfɪkənt] *adj.* 重要的
engagement [ɪn'geɪdʒmənt] *n.* 约定；保证
virtualize ['vɜ:tʃʊəlaɪz] *v.* 使虚拟化
vulnerability [ˌvʌlnərə'bɪləti] *n.* 脆弱点；脆弱性

They have a significant physical footprint, requiring allocations of space, hardware, and environmental controls. These **assets** have to be **refreshed periodically**, resulting in additional capital **expenditures**. They have attracted **criticism** because users "still have to buy, build, and manage them" and thus do not benefit from less hands-on management, **essentially** "[lacking] the economic model that makes cloud computing such an **intriguing** concept".

Public cloud

A cloud is called a "public cloud" when the services are **rendered** over a network that is open for public use. Public cloud services may be free. Technically there may be little or no difference between public and private cloud architecture. However, security consideration may be substantially different for services (applications, storage, and other resources) that are made available by a service provider for a public audience and when communication is effected over a non-trusted network. Generally, public cloud service providers like Amazon Web Services (AWS), IBM Cloud, Oracle, Microsoft, Google, and Alibaba own and operate the infrastructure at their data center and access is generally via the Internet. AWS, Oracle, Microsoft, and Google also offer direct connect services called "AWS Direct Connect", "Oracle FastConnect", "Azure ExpressRoute", and "Cloud Interconnect" respectively. Such connections require customers to purchase or lease a private connection to a peering point offered by the cloud provider.

Hybrid cloud

Hybrid cloud is a composition of a public cloud and a private environment, such as a private cloud or on-premises resources, that remain distinct entities but are bound together, offering the benefits of multiple deployment models. Hybrid cloud can also mean the ability to connect **collocation**, managed and/or dedicated services with cloud resources. Gartner defines a hybrid cloud service as a cloud computing service that is composed of some combination of private, public and community cloud services, from different service providers. A hybrid cloud service crosses isolation and provider boundaries so that it can't be simply put in one **category** of private, public, or community cloud service. It allows one to extend either the capacity or the capability of a cloud service, by **aggregation**, **integration** or **customization** with another cloud service.

Varied use cases for hybrid cloud composition exist. For example, an organization may store sensitive client data in house on a private cloud application, but interconnect that application to a business intelligence application provided on a public cloud as a software

asset ['æset] *n.* 资产；财产
refresh [rɪ'freʃ] *v.* 使恢复精神
periodically [ˌpɪərɪ'ɒdɪkli] *adv.* 定期地；周期性地
expenditure [ɪk'spendɪtʃə(r)] *n.* 花费
criticism ['krɪtɪsɪzəm] *n.* 批评；批判
essentially [ɪ'senʃəli] *adv.* 基本上；本质上
intriguing [ɪn'tri:gɪŋ] *adj.* 吸引人的；有趣的
render ['rendə(r)] *v.* 给予；提供

hybrid ['haɪbrɪd] *n.* 混合物；*adj.* 混合的
hybrid cloud 混合云
premise ['premɪs] *n.* 建筑物及其土地；生产场所
distinct [dɪ'stɪŋkt] *adj.* 可辨别的
collocation [ˌkɒlə'keɪʃn] *n.* 搭配
category ['kætəgəri] *n.* 种类；类别
aggregation [ˌægrɪ'geɪʃn] *n.* 聚合；集合体
integration [ˌɪntɪ'greɪʃn] *n.* 整合
customization ['kʌstəmaɪzeɪʃən] *n.* 定制；[计]用户化

service. This example of hybrid cloud extends the capabilities of the enterprise to deliver a specific business service through the addition of externally available public cloud services. Hybrid cloud adoption depends on a number of factors such as data security and **compliance** requirements, levels of control needed over data, and the applications an organization uses.

compliance [kəm'plaɪəns] *n.* 服从；遵从

Another example of hybrid cloud is one where IT organizations use public cloud computing resources to meet temporary capacity needs that cannot be met by the private cloud. This capability enables hybrid clouds to employ cloud **bursting** for scaling across clouds. Cloud bursting is an application **deployment** model in which an application runs in a private cloud or data center and "bursts" to a public cloud when the demand for computing capacity increases. A primary advantage of cloud bursting and a hybrid cloud model is that an organization pays for extra compute resources only when they are needed. Cloud bursting enables data centers to create an in-house IT infrastructure that supports average workloads, and use cloud resources from public or private clouds, during spikes in processing demands. The specialized model of hybrid cloud, which is built atop **heterogeneous** hardware, is called "cross-platform hybrid cloud". A cross-platform hybrid cloud is usually powered by different CPU architectures, underneath for example, x86-64 and ARM. Users can **transparently** deploy and scale applications without knowledge of the cloud's hardware diversity. This kind of cloud emerges from the rise of ARM-based system-on-chip for server-class computing.

burst [bɜːst] *v.* 爆炸

deployment [dɪ'plɔɪmənt] *n.* 部署；调度

heterogeneous [ˌhetərə'dʒiːniəs] *adj.* 不同的

transparently [træns'pærəntli] *adv.* 透明地；明显地

.End.

cloud computing 云计算	predominant *adj.* 占优势的
enterprise cloud 企业云	public cloud 公共云
popularize *vt.* 普及	predecessor *n.* 前辈；前任；前身
denote *v.* 表示	schematic *n.* 图表
simplification *n.* 简化	implication *n.* 暗示
endpoint *n.* 端点	diagram *n.* 图解
private cloud 私有云	infrastructure *n.* 基础结构
undertake *v.* 保证；任务	significant *adj.* 重要的
engagement *n.* 约定；保证	virtualize *v.* 使虚拟化
vulnerability *n.* 脆弱点；脆弱性	asset *n.* 资产；财产

Key Words

refresh *v.* 使恢复精神
expenditure *n.* 花费
essentially *adv.* 基本上；本质上
render *v.* 给予；提供
hybrid cloud 混合云
distinct *adj.* 可辨别的
category *n.* 种类；类别
integration *n.* 整合
compliance *n.* 服从；遵从
deployment *n.* 部署
transparently *adv.* 透明地；明显地
periodically *adv.* 定期地；周期性地
criticism *n.* 批评；批判
intriguing *adj.* 吸引人的；有趣的
hybrid *n.* 混合物
premise *n.* 建筑物及其土地；生产场所
collocation *n.* 搭配
aggregation *n.* 聚合；集合体
customization *n.* 定制；[计]用户化
burst *v.* 爆炸
heterogeneous *adj.* 不同的

参考译文 技能2 云计算

云计算是计算机系统资源[尤其是数据存储（云存储）和计算能力]的按需可用性，而无须用户直接进行主动管理。该术语通常用于描述Internet上可供许多用户使用的数据中心。如今占主导地位的大型云通常具有从中央服务器分布在多个位置的功能。如果与用户的连接相对紧密，则可以将其指定为边缘服务器。

云可能仅限于一个组织（企业云），也可能对许多组织可用（公共云）。

云计算依靠资源共享来实现一致性和规模经济。

历史

随着Amazon.com在2006年发布其Elastic Compute Cloud产品，云计算得到了普及。

早在1996年，“云计算”一词就出现在康柏的内部文件中。

早在1977年，在最初的ARPANET和1981年的CSNET中，云符号就被用来代表计算机设备的网络-它们都是互联网的前身。

“云”一词被用作互联网的隐喻，而标准化的类似于云的形状被用来表示电话原理图上的网络。通过这种简化，这意味着网络端点如何连接的细节与理解该图无关。

早在1993年，“云”一词就曾指代分布式计算的平台，当时苹果公司的分公司General Magic和AT&T用它来描述他们的（配对的）Telescript和PersonaLink技术。在Wired1994年4月的专题片《比尔和安迪的奇妙冒险II》中，安迪•赫兹菲尔德（Andy Hertzfeld）对General Magic的分布式编程语言Telescript进行了评论：

“Telescript的美丽之处在于……现在，我们不仅拥有一个编程的设备，而且我们拥有了整个云，在那里，一个程序就可以运行并传播到许多不同的信息源，并创建一种虚拟服务，以前没有人想到过，Jim White [Telescript，X.400和ASN.1的设计者现在使用的示例是日期安排服务，其中软件代理去花店订购鲜花，然后去售票处并获得演出门票，一切都传达给了双方。”

云计算类型

私有云

私有云是专门为单个组织运营的云基础架构，无论是内部管理还是第三方管理，都可以在内部或外部托管。进行私有云项目需要大量参与以虚拟化业务环境，并且需要组织重新评估有关现有资源的决策。它可以改善业务，但是项目的每个步骤都会引发安全问题，必须解决这些问题才能防止严重的漏洞。自主运营数据中心通常是资本密集型的。它们占用大量物理空间，需要分配空间，硬件和环境控制。这些资产必须定期更新，从而导致额外的资本支出。他们之所以招致外界批评，是因为用户“仍然必须购买、构建和管理它们”，因此，无法从较少的实际操作管理中获益，本质上是“（缺乏）使云计算成为如此吸引人的概念的经济模式”。

公共云

当通过开放供公众使用的网络提供服务时，云被称为“公共云”。公共云服务可能是免费的。从技术上讲，公共云和私有云体系结构之间可能几乎没有差异，但是，对于服务提供商向公众提供的服务（应用程序，存储和其他资源），通过不可信网络进行通信时，安全性考虑可能有很大不同。通常，诸如Amazon Web Services（AWS），IBM Cloud，Oracle，Microsoft，Google和Alibaba的公共云服务提供商在其各自的网站上拥有并运营基础架构数据中心和访问通常是通过互联网进行的。AWS，Oracle，Microsoft和Google还分别提供称为“AWS Direct Connect”，“Oracle FastConnect”，“ Azure ExpressRoute”和“ Cloud Interconnect”的直接连接服务，此类连接需要客户购买或租用专用连接到云提供商提供的对等点。

混合云

混合云是由公共云和私有环境（例如私有云或本地资源）组成的，它们仍然是不同的实体，但被捆绑在一起，提供了多种部署模型的优势。混合云还意味着能够将搭配服务，托管服务和/或专用服务与云资源进行连接。Gartner将混合云服务定义为一种云计算服务，由来自不同服务提供商的私有，公共和社区云服务的某种组合组成。混合云服务跨越了隔离和提供商边界，因此不能简单地将其归为一类私有、公共或社区云服务。它允许用户通过与另一种云服务的聚合，集成或定制来扩展云服务的容量或能力。

存在各种混合云组成的用例。例如，组织可以将敏感的客户端数据存储在私有云应用程序的内部，但可以将该应用程序与作为软件服务在公共云上提供的商业智能应用程序互连。混合云的此示例通过添加外部可用的公共云服务扩展了企业提供特定业务服务的功能。混合云的采用取决于许多因素，例如数据安全性和合规性要求，对数据所需的控制级别以及组织使用的应用程序。

混合云的另一个示例是IT组织使用公共云计算资源来满足私有云无法满足的临时容量需求的示例。此功能使混合云能够使用云爆发来跨云扩展。云爆发是一种应用程序部署模型，在该模型中，应用程序在私有云或数据中心中运行，并在对计算能力的需求增加时“爆发”到公共云。云爆发和混合云模型的主要优势在于，组织仅在需要时才为额外的计算资源付费。云爆发使数据中心能够创建支持平均工作负载的内部IT基础架构，并在处理需求激增期间使用来自公共或私有云的云资源。建立在异构硬件之上的混合云的专用模型称为“跨平台混合云”。跨平台的混合云通常由下面的不同CPU架构提供支持，例如x86-64和ARM。用户可以明显地部署和扩展应用程序，而无须了解云的硬件多样性。这种云是基于ARM用于服务器级计算的系统芯片的兴起而产生的。

Fast Reading One Information Retrieval

Information retrieval (IR) is the activity of obtaining information system resources that are relevant to an information need from a collection of those resources. Searches can be based on full-text or other content-based indexing. Information retrieval is the science of searching for information in a document, searching for documents themselves, and also searching for the metadata that describes data, and for databases of texts, images or sounds.

Fast Reading One

Automated information retrieval systems are used to reduce what has been called information overload. An IR system is a software system that provides access to books, journals and other documents; stores and manages those documents. Web search engines are the most visible IR applications.

Overview

An information retrieval process begins when a user enters a query into the system. Queries are formal statements of information needs, such as search strings in web search engines. In information retrieval a query does not uniquely identify a single object in the collection. Instead, several objects may match the query, perhaps with different degrees of relevancy.

An object is an entity that is represented by information in a content collection or database. User queries are matched against the database information. However, as opposed to classical SQL queries of a database, in information retrieval the results returned may or may not match the query, so results are typically ranked. This ranking of results is a key difference of information retrieval searching compared to database searching.

Depending on the application the data objects may be, for example, text documents, images, audio, mind maps or videos. Often the documents themselves are not kept or stored directly in the IR system, but are instead represented in the system by document surrogates or metadata.

Most IR systems compute a numeric score on how well each object in the database matches the query, and rank the objects according to this value. The top ranking objects are then shown to the user. The process may then be iterated if the user wishes to refine the query.

History

The idea of using computers to search for relevant pieces of information was popularized in the article *As We May Think* by Vannevar Bush in 1945. It would appear that Bush was inspired by patents for a 'statistical machine' — filed by Emanuel Goldberg in the 1920s and 1930s — that searched for documents stored on film. The first description of a computer searching for information was described by Holmstrom in 1948, detailing an early mention of the Univac computer. Automated information retrieval systems were introduced in the 1950s: one even featured in the 1957 romantic comedy, Desk Set. In the 1960s, the first large information retrieval research group was formed by Gerard Salton at Cornell. By the 1970s several different retrieval techniques had been shown to perform well on small text corpora such as the Cranfield collection (several thousand documents). Large-scale retrieval systems, such as the Lockheed Dialog system, came into use early in the 1970s.

In 1992, the US Department of Defense along with the National Institute of Standards and Technology (NIST), cosponsored the Text Retrieval Conference (TREC) as part of the TIPSTER text program. The aim of this was to look into the information retrieval community by supplying the infrastructure that was needed for evaluation of text retrieval methodologies on a very large text collection. This catalyzed research on methods that scale to huge corpora.

The introduction of web search engines has boosted the need for very large scale retrieval systems even further.

.End.

参考译文 快速阅读1 信息检索

信息检索（IR）是从这些资源的集合中获取与信息需求相关的信息系统资源的活动。搜索可以基于全文索引或其他基于内容的索引。信息检索是在文档中搜索信息，搜索文档本身，还搜索描述数据的元数据以及文本，图像或声音的数据库的科学。

自动化的信息检索系统用于减少所谓的信息过载。IR系统是一种软件系统，可以访问书籍，期刊和其他文档。存储和管理这些文档。Web搜索引擎是最可见的IR应用程序。

概述

当用户向系统中输入查询时，信息检索过程开始。查询是信息需求的正式声明，例如Web搜索引擎中的搜索字符串。在信息检索中，查询不会唯一地标识集合中的单个对象。取而代之的是，几个对象可能与查询匹配，也许具有不同的相关度。

对象是由内容集合或数据库中的信息表示的实体。用户查询与数据库信息匹配。但是，与数据库的经典SQL查询相反，在信息检索中，返回的结果可能匹配查询，也可能不匹配，因此通常对结果进行排名。这种排名结果是相比于数据库检索信息检索搜索的关键区别。

取决于应用，数据对象可以是例如文本文档，图像，音频，思维导图或视频。通常，文档本身不会保存或直接存储在IR系统中，而是由文档替代或元数据在系统中表示。

大多数IR系统都会对数据库中每个对象与查询的匹配程度进行数字评分，然后根据该值对对象进行排名。然后将排名最高的对象显示给用户。如果用户希望优化查询，则可以重复该过程。

历史

1945年，范内瓦尔•布什（Vannevar Bush）在《我们可能会想》一文中普及了使用计算机搜索相关信息的想法。看来，布什受到了“统计机器”专利的启发-由伊曼纽尔•戈德堡（Emanuel Goldberg）申请在20世纪20年代和30年代搜索电影中存储的文档。1948年，霍姆斯特罗姆（Holmstrom）对计算机搜索信息进行了首次描述，较早提到了Univac计算机。20世纪50年代引入了自动信息检索系统：甚至在1957年的浪漫喜剧《书桌套装》中也采用了这种系统。1960年代，康奈尔大学的杰拉德•萨尔顿（Gerard Salton）成立了第一个大型信息检索研究小组。到20世纪70年代，已经显示出几种不同的检索技术在小型文本语料库（例如Cranfield集合）（数千份文档）中表现良好。大型检索系统，例如洛克希德对话系统，于20世纪70年代初投入使用。

1992年，美国国防部与美国国家标准技术研究院（NIST）共同发起了文本检索会议（TREC），作为TIPSTER文本程序的一部分。其目的是通过提供评估非常大的文本集上的文本检索方法所需要的基础结构来调查信息检索社区。这促进了对可扩展到大型语料库的方法的研究。网络搜索引擎的引入进一步提高了对超大规模检索系统的需求。

Fast Reading Two The Structure of BigData

Big data repositories have existed in many forms, often built by corporations with a special need. Commercial vendors historically offered parallel database management systems for big data beginning in the 1990s. For many years, WinterCorp published the largest database report.

Teradata Corporation in 1984 marketed the parallel processing DBC 1012 system. Teradata systems were the first to store and analyze 1 terabyte of data in 1992. Hard disk drives were 2.5 GB in 1991 so the definition of big data continuously evolves according to Kryder's Law. Teradata installed the first petabyte class RDBMS based system in 2007. As of 2017, there are a few dozen petabyte class Teradata relational databases installed, the largest of which exceeds 50 PB. Systems up until 2008 were 100% structured relational data. Since then, Teradata has added unstructured data types including XML, JSON, and Avro.

Fast Reading Two

In 2000, Seisint Inc. (now LexisNexis Risk Solutions) developed a C++based distributed platform for data processing and querying known as the HPCC Systems platform. This system automatically partitions, distributes, stores and delivers structured, semi-structured, and unstructured data across multiple commodity servers. Users can write data processing pipelines and queries in a declarative dataflow programming language called ECL. Data analysts working in ECL are not required to define data schemas upfront and can rather focus on the particular problem at hand, reshaping data in the best possible manner as they develop the solution. In 2004, LexisNexis acquired Seisint Inc. and their high-speed parallel processing platform and successfully utilized this platform to integrate the data systems of Choicepoint Inc. when they acquired that company in 2008. In 2011, the HPCC systems platform was open-sourced under the Apache v2.0 License.

CERN and other physics experiments have collected big data sets for many decades, usually analyzed via high-throughput computing rather than the map-reduce architectures usually meant by the current "big data" movement.

In 2004, Google published a paper on a process called MapReduce that uses a similar architecture. The MapReduce concept provides a parallel processing model, and an associated implementation was released to process huge amounts of data. With MapReduce, queries are split and distributed across parallel nodes and processed in parallel (the Map step). The results are then gathered and delivered (the Reduce step). The framework was very successful, so others wanted to replicate the algorithm. Therefore, an implementation of the MapReduce framework was adopted by an Apache open-source project named Hadoop. Apache Spark was developed in 2012 in response to limitations in the MapReduce paradigm, as it adds the ability to set up many operations (not just map followed by reducing).

MIKE2.0 is an open approach to information management that acknowledges the need for revisions due to big data implications identified in an article titled "Big Data Solution Offering". The methodology addresses handling big data in terms of useful permutations of data sources, complexity in interrelationships, and difficulty in deleting (or modifying) individual records.

2012 studies showed that a multiple-layer architecture is one option to address the issues that big data presents. A distributed parallel architecture distributes data across multiple servers; these parallel execution environments can dramatically improve data processing speeds. This type of architecture inserts data into a parallel DBMS, which implements the use of MapReduce and Hadoop frameworks. This type of framework looks to make the processing power transparent to the end-user by using a front-end application server.

The data lake allows an organization to shift its focus from centralized control to a shared model to respond to the changing dynamics of information management. This enables quick segregation of data into the data lake, thereby reducing the overhead time.

.End.

参考译文 快速阅读2 大数据的架构

大数据存储库以多种形式存在，通常由有特殊需要的公司构建。从20世纪90年代开始，商业供应商就一直为大数据提供并行数据库管理系统。多年来，WinterCorp发布了最大的数据库报告。

Teradata Corporation在1984年销售了并行处理DBC 1012系统。Teradata系统是1992年第一个存储和分析1 TB数据的系统。1991年，硬盘驱动器为2.5 GB，因此大数据的定义根据Kryder定律不断发展。Teradata在2007年安装了第一个基于PB级RDBMS的系统。截至2017年，已安装了数十个PB级Teradata关系数据库，其中最大的数据库超过50PB。截至2008年，系统都是100%结构化的关系数据。从那时起，Teradata已添加非结构化数据类型，包括XML，JSON和Avro。

2000年，Seisint Inc.（现为LexisNexis Risk Solutions）开发了一种用于数据处理和查询的基于C++的分布式平台，称为HPCC系统平台。该系统跨多个商品服务器自动分区、分发、存储和交付结构化、半结构化和非结构化数据。用户可以使用被称为ECL的声明性数据流编程语言编写数据处理管道和查询。在ECL中工作的数据分析人员不需要预先定义数据模式，而是可以专注于手头的特定问题，在开发解决方案时以最佳方式重塑数据。2004年，LexisNexis收购了Seisint Inc.及其高速并行处理平台，并在2008年收购Choicepoint Inc.时成功利用了该平台来整合这类公司的数据系统。2011年，HPCC系统平台在Apache v2.0许可下开放源代码。

欧洲核子研究组织（CERN）和其他物理实验已经收集了数十年的大数据集，通常通过高通量计算进行分析，而不是通过当前“大数据”运动通常意指的地图缩减架构来进行分析。

2004年，Google发表了一篇关于使用类似架构的名为MapReduce的过程的论文。MapReduce概念提供了并行处理模型，并发布了相关的实现以处理大量数据。使用MapReduce，查询可以被拆分并分布在多个并行节点上，并可以被并行处理（映射步骤）。然后收集并交付结果（缩小步骤）。该框架非常成功，因此其他人想复制该算法。因此，Apache开源项目Hadoop采纳了MapReduce框架的实现。Apache Spark是针对MapReduce范式的局限性于2012年被开发的，因为它增加了设置许多操作（不仅是映射再缩小）的功能。

MIKE2.0是一种开放的信息管理方法，该方法承认由于标题为“大数据解决方案产品”的文章中提到的大数据影响而需要进行修订。该方法论根据数据源的有用排列，相互关系的复杂性以及删除（或修改）单个记录的难度来处理大数据。

2012年的研究表明，多层体系结构是解决大数据存在的问题的一种选择。分布式并行体系结构分布在多个服务器上的数据；这些并行执行环境可以大大提高数据处理速度。这种类型的体系结构将数据插入到并行DBMS中，该DBMS实现了MapReduce和Hadoop框架的使用。这种类型的框架通过使用前端应用程序服务器来使处理能力对最终用户透明。

数据湖允许组织将其焦点从集中控制转移到共享模式，以响应信息管理的动态变化。

Ex 1 What is big data? Try to give a brief summary according to the passage.

Ex 2 Fill in the table below by matching the corresponding Chinese or English equivalents.

big data	
	行
veracity	
	仿真
public cloud	
	混合云

Ex 3 Choose the best answer to each of the following statements according to the text we've learnt.

1. Big data was originally associated with three key concepts: volume, variety, and __________.
   A. interfaces　　B. programs
   C. velocity　　D. account numbers
2. The big data has been in use since the ____________.
   A. 2000s　　B. 1990s
   C. 1800s　　D. 1890s
3. Big data can be described by the following characteristics ____________.
   A. Windows　　B. veracity
   C. data　　D. cloud
4. If the connection to the user is relatively close, it may be designated by _____________.
   A. an edge server　　B. Plug and Play
   C. Media Player　　D. Windows 95
5. Which of the following is one of the Cloud Computing models ____________?
   A. Windows　　B. Unix
   C. hybrid cloud　　D. 3ds Max

Part B Practical Learning

Task One

Task Two

Training Target

In this part, there are two tasks in English environment. You should complete these tasks in groups under the joint guidance of professional teachers and laboratory teachers, so as to train and improve your ability to complete professional tasks in English environment.

Task One Inquire Information on the Internet

The first task is to inquire the necessary information on the Internet. In this task, you must know the necessary information needed to finish the whole task.

There are some information about the necessary information on the Internet. The information can help you to finish the task.

The needed information can be downloaded through the webpage: www. baidu.com or www. google.com. For example (Pic 4.1, Pic 4.2).

Pic 4.1 baidu

Pic 4.2 google

You can click on the relevant provisions to enter into relevant pages for details. (Pic 4.3, Pic 4.4)

Pic 4.3 Page for details (1)

Pic 4.4 Page for details (2)

Task Two Choose Useful Information

In this task, students choose most useful information from what they are looking for (Pic 4.5)

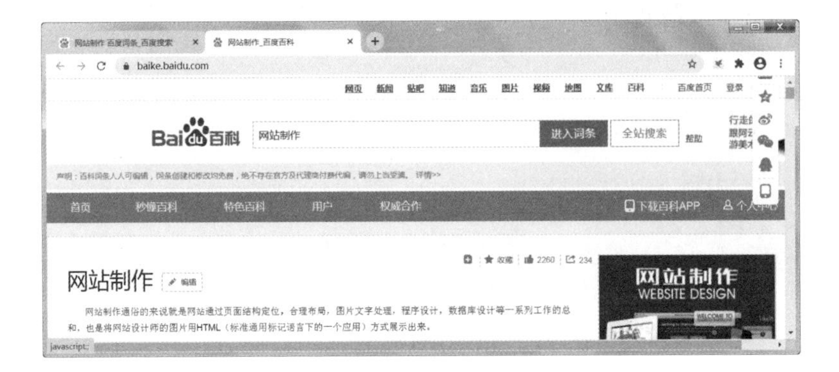

Pic 4.5 choose useful information

Part C Occupation English

Occupation English

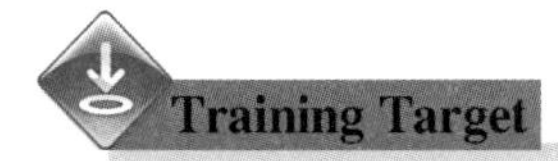

In this part, there is an English dialogue in real life and work environment. You will play the roles of A and B and read the dialogue aloud to practice your ability to use English.

How to Search Information Efficiently

如何高效地搜索信息

Role Setting: Student (A), Student (B)

角色设置：学生（A），学生（B）

A: I recently found that searching information online is also a technical activity. 我最近发现，在网上搜索信息也是个技术活。

B: Of course, there are sea information resources on the Internet. How to quickly and accurately find what you want needs some knowledge. 那当然了，网上有海量的信息资源，如何快速、准确地找到你想要的是需要点学问的。

A: Listen to you, do you have any unique skills? 听你说的，难道你有什么绝招？

B: I can't talk about the unique skills, but I still have some experience. 绝招谈不上，心得还是有的。

A: How to effectively achieve information search? I think there are a few points to be clear. 如何高效地实现信息搜索，我觉得有几点要明确。

B: Don't beat around the bush. Talk about it. 别绕弯子了，快说来听听。

A: Before searching, you should be clear about three things: what to search, that is, confirm the subject of the search; where to search, select the appropriate resource library; select information, and use filtering rules and sorting rules to select effective information. 搜索前，你要明确三件事：搜什么，即确认搜索的主题；在哪搜，选择合适的资源库；选信息，使用筛选规则和排序规则挑选有效信息。

B: Yes, I think the second and third steps are easy to determine in these three steps. The first step is to think about the topic. This step will save a lot of time. 是的，我觉得在这三步里第二、第三步都好确定。第一步要好好考虑主题，这一步（确定做好了）会节省很多时间。

A: Of course. Search is to solve a problem, but it may not be solved by a single question, so is search, which may require a series of searches to find the answer. 那当然了。搜索是为了解决一个问题，但解决问题可能不是通过一次提问就能搞定的，搜索也是如此，可能需要进行一系列搜索才能发现答案。First, learn to break down the problem you are searching for. Blind "direct search" is likely to be futile, and it may be more efficient to refine the steps of separating components for the problems to be searched. Second, describe exactly the question you want to search for, and finally use the right keywords. If these three steps are done well, the efficiency of search will be greatly improved. 首

先，学会分解你要搜索的问题。盲目地“直接搜索”很可能徒劳无功，将所搜寻的问题细分为几个步骤的小问题可能更加高效。其次，精确描述你要搜索的问题，最后用对关键词。如果将这三步做好了，搜索的效率会大大提高。

B: Is that reasonable? I'll try it right away. 很有道理吗？我要立即试试。

A: Good luck. 祝你好运。

Word Building

前缀/后缀由一个或几个字母组成，放在词根或单词之前/之后，组成一个新词。

（1）dis-（前缀）:表示“不,消失掉”

like 喜欢 ———— dislike 不喜欢

order 顺序 ———— disorder 无秩序

honest 诚实的 ———— dishonest 不诚实的

（2）im-（前缀）：表示“不，无，非”

Possible 可能的 ———— impossible 不可能的

moral 有道德的 ———— immoral 不道德的

polite 礼貌的 ———— impolite 无礼貌的

（3）ful（后缀）：表示“有……的”

grate 感激 ———— grateful 感激的

rue 悔恨 ———— rueful 后悔的

will 意志 ———— willful 任性的

（4）-less（后缀）:表示“无……的,不……的”

spine 脊骨 ———— spineless 没有骨气的

nerve 勇气；神经 ———— nerveless 无勇气的

feck 效果 ———— feckless 无计划的

Ex Translate the following words and try your best to guess the meaning of each word on the right according to the clues given on the left.

tire	疲倦（动词）	tireless ______
home	家（名词）	homeless ______
color	颜色（名词）	colorless ______
partial	有偏见的（形容词）	impartial ______
mortal	不能永生的（形容词）	immortal ______
passionate	有激情的（形容词）	dispassionate ______
appear	出现（动词）	disappear ______
arm	武装（动词）	disarm ______
cover	盖（动词）	discover ______
care	照顾（动词）	careful ______
wonder	惊讶（名词）	wonderful ______
color	颜色（名词）	colorful ______

Ex 1 What is cloud computing?

Ex 2 What is public cloud? Try to give several examples.

Ex 3 What are the big data characteristics of?

Ex 4 Fill in the table below by matching the corresponding Chinese or English equivalents.

cloud computing	
	大数据
enterprise cloud	
	速度
public cloud	
	网页
variety	
	混合云
virtualize	

Ex 5 Choose the best answer to each of the following statements according to the text we've learnt.

1. Cloud computing was popularized with releasing ____________ its Elastic Compute Cloud product in 2006.
   A. Amazon.com
   B. Google
   C. Baidu
   D. Microsoft
2. Big data has been in use since the 1990s, with some giving credit to ____________ for popularizing the term.
   A. Ma Yun
   B. John Mashey
   C. Tom
   D. Bill Gates
3. A cloud is called a " ____________ " when the services are rendered over a network that is open for public use.
   A. operation system
   B. public cloud
   C. office software

D. application software

4. Compared to small data, ___________ are (is) produced more continually.

A. word

B. data

C. big data

D. DBMS

5. Hybrid cloud is a composition of a public cloud and a ______________.

A. software

B. data

C. private environment

D. A and B

Project Five

Designing Online Stores

Part A Theoretical Learning

Part B Practical Learning

Part C Occupation English

Part A Theoretical Learning

Training Target

In this part, our target is to improve the speed of reading professional articles and the comprehension ability of the reader. We have marked specialized key words and some flexible sentences. Try to grasp the main idea of each paragraph.

Skill One A Short Introduction to the Internet

Internet is a giant global and open computer network, which is a collection of interconnected networks. It is also a way to exchange information and share resource between **international** computers.

The original Internet is ARPANET(Advanced Research Projects Agency Network) , which was established by U.S. Department of Defense in 1969. The ARPANET was used to share data between some separated **military** institutes and universities in region. By 1972 the network had expanded to incorporate 40 **nodes**. Many U.S. government agency networks had been linked by ARPANET, and because the networks were of a disparate nature, a common network **protocol** called TCP/IP (Transmission Control Protocol/Internet Protocol) was developed and became the standard for Internet working military computers.

When the U.S. was developing their national net, the other countries were developing theirs, too. At the end of the 1980s, interlinkage of different countries' computer net appears. After that there were countries joining in every year and getting to form the present Internet. It is growing so quickly that nobody can say exactly how many users are "On the Net" now.

Once your computer makes a connection with the Internet, you will find that you have walked into the largest **repository** of information — a magic world. The following are the important service functions that the Internet provides.

1. E-mail

The most widely used tool on the Internet is electronic mail or E-mail. E-mail enables you to send messages to Russia, Japan and so on, no matter how far between individuals. E-mail messages are

interconnect [ˌɪntəkə'nekt] *v.* 互联

international [ˌɪntə'næʃnəl] *adj.* 国际的

military ['mɪlətri] *adj.* 军事的，军用的

node [nəʊd] *n.* 节点

protocol ['prəʊtəkɒl] *n.* 协议

Skill One

repository [rɪ'pɒzətri] *n.* 仓库，资源丰富的地方

generally sent from and received by **mail servers** — computers that are **dedicated to** processing and directing E-mail. As a very convenient and inexpensive way to transmit messages, E-mail has dramatically affected scientific, personal and business communications. In some cases, E-mail has replaced the telephone for conveying messages.

mail server 邮件服务器
dedicate…to 把……用于

2. File Transfer Protocol

File Transfer Protocol (FTP) is a method of transferring files from one computer to another over the Internet, even if each computer has a different operating system or storage format. FTP is designed to **download** files (e.g. receive from the Internet) or **upload** files (e.g. send to the Internet). The ability to upload and download files on it is one of the most valuable features the Internet offers. This is especially helpful for those people who rely on computers for various purposes and may need software drivers and **upgrades** immediately. Network administrators can rarely wait even a few days to get the necessary drivers that enable their network severs to function again. The Internet can provide these files immediately.

download [daʊn'ləʊd] *v.* 下载
upload [ˌʌp'ləʊd] *v.* 上传
upgrade [ˌʌp'greɪd] *n.* 升级，软件升级

3. The World Wide Web

The World Wide Web (WWW), which Hypertext Transfer Protocol (HTTP) works with, is the fastest growing and most widely-used part of the Internet. The WWW is a way to exchange information between computers and the Internet. **Hyperlink** makes the Internet easy to navigate. It is an object (word, phrase, or picture) on a webpage that, when clicked, transfers you to a new webpage. One of the main reasons for the extraordinary growth of the web is the ease in which it allows access to information. One limitation of HTTP is that you can only use it to download files, but you cannot upload them.

hypertext ['haɪpətekst] *n.* 超文本
hyperlink ['haɪpəlɪŋk] *n.* 超链接

4. Telnet

Telnet allows an Internet user to connect to a distance computer and use that computer as if he or she was using it directly. To make a connection with a telnet client, you must select a connection option: "Host Name" and "**Terminal** Type". The host name is the IP address (**DNS**) of the remote computer to which you connect. The terminal type describes the type of terminal **emulation** that you want the computer to perform.

The Internet has many new technologies, such as global chat, **video conferencing**, free international phone and more. The Internet becomes more and more popular in society in recent years. So we can say that the Internet is your PC's window to the rest of the world.

telnet ['telnet] *n.* 远程登录
terminal ['tɜ:mɪnl] *n.* 终端
DNS (Domain Name System) 域名服务器
emulation [ˌemju'leɪʃn] *n.* 竞争，仿真
video conferencing 视频会议

.End.

Key Words

interconnect *v.* 互联
military *adj.* 军事的，军用的
protocol *n.* 协议
mail server 邮件服务器
download *v.* 下载
upgrade *n.* 升级，软件升级
hyperlink *n.* 超链接
DNS 域名服务器
video conferencing 视频会议
international *adj.* 国际的
node *n.* 节点
repository *n.* 仓库，资源丰富的地方
dedicate. . . to 把……用于
upload *v.* 上传
hypertext *n.* 超文本
telnet *n.* 远程登录
emulation *n.* 竞争，仿真

参考译文 技能1 互联网简介

互联网是一个巨大的全球性开放式计算机网络，它由众多网络互联而成。它也是全球计算机之间实现国际信息交流和资源共享的一种方式。

互联网的前身是美国国防部于1969年建立的ARPANET（高级项目研究机构网络），它是为了让在地域上相互分离的一些军事研究机构和大学之间实现数据共享而开发的。到1972年，这个网络扩展到可以包含40个节点。美国许多政府网络也连入了ARPANET，由于网络的不同，一个被称为TCP/IP(传输控制协议和网际协议)的网络协议被发展起来，成为网络中的军用计算机都要遵守的标准协议。

在美国发展自己的网络的同时,其他国家也在发展自己的网络。到20世纪80年代末，不同国家的计算机网络连接出现了。从那以后，每年都有一些国家加入其中，逐步形成了现在的互联网。互联网发展如此之快以至于没有人能准确地说出网上到底有多少用户。

一旦你的计算机连入了互联网，你就会发现你进入了一个最大的信息宝库——一个神奇的世界。下面是互联网提供的重要服务功能。

1. 电子邮件

在互联网上应用最广泛的工具是电子邮件，简称E-mail。通过电子邮件，你可以发送信息到俄罗斯、日本等，无论两人之间的距离有多远。通常都是由邮件服务器负责发送和接收邮件信息，邮件服务器是用来处理和传送邮件的计算机。利用电子邮件传递信息是非常方便而且便宜的方式，电子邮件已经极大地影响了科学、个人以及企业之间的通信。有些时候，电子邮件在信息传递上已经替代了电话。

2. 文件传输协议

文件传输协议（FTP）是一台计算机和另一台计算机之间通过互联网进行文件传输的一种方法，即使各个计算机有着不同的操作系统和存储格式。FTP可以用来下载文件（例如从互联网上获取）或上传文件（例如发送到互联网上）。上传和下载文件的功能是互联网所提供的最有价值的功能之一。这些功能（上传和下载）对于那些经常利用计算机实现各种应用，以及常常需要各类软件驱动程序且随时需要升级软件的人来说非常有帮助。当网络管理者急需一些驱动程序来使网络服务器重新启动时，哪怕是几天的时间他们也是等不了的。通过使用FTP，互联网能够迅速提供这些文件。

3．万维网

利用超文本传输协议（HTTP）的万维网（WWW）是互联网中发展最快、应用最广的部分。万维网是实现计算机与互联网之间信息传递的一种手段。超链接使我们漫游互联网更加容易。超链接指的是网页上的一些文字、短语、图片等，当单击它时，它会带你进入一个新的网页。网络飞速发展的主要原因是它让人们很容易获取信息。HTTP的一个局限性是它只能用来下载文件而不能上传文件。

4．远程登录

远程登录使互联网的用户能够登录到远程的计算机上，而且就像他（或她）亲自在使用那台计算机一样。要想登录到远程计算机上，你必须对主机名和终端类型进行选择。主机名就是你所要连接的计算机的IP地址（域名服务器），终端类型指的是你所希望计算机扮演的终端仿真的类型。

互联网还包含许多新的技术，如全球聊天、视频会议、免费国际电话等。互联网近几年在社会上的影响越来越大，所以我们可以说，互联网是你的PC机通向世界其他地方的窗口。

Skill Two Website Design

Many people wish to create a flashy **website**. But creating a great website doesn’t happen at the tips of the fingers; it happens in the depths of the brain. Outstanding websites result from extensive planning. **Prior** preparation saves time and avoids **frustration** both during page creation and when updates and additions are required. The three-step design **tutorial** will show you how to create a high-end attractive website.

website ['websaɪt] *n.* 网站

prior ['praɪə(r)] *adj.* 在先的，优先的

frustration [frʌ'streɪʃn] *n.* 挫折，挫败

tutorial [tju:'tɔ:riəl] *n.* 辅导，指导

Step One—Determining Who Use Your Site and Their Information Needs

Successful websites know who their customers are and why they visit, and they provide a responsive and attractive display to those viewers. Customers don’t visit our site because we spend time creating it; customers deserve maximum benefit from the time they **allocate** to us.

allocate ['æləkeɪt] *v.* 分配

Step Two—Editing Your Webpage

1. Establish an **identity** and use it consistently on all pages.

Viewers of our webpages should know exactly who we are, and after linking should know if they're still on one of our site's pages. It doesn't mean every page looks the same, but the colors and graphics we use should be consistent throughout the website. Establish a **theme** or identifying characteristic for your website.

2. Create user-friendly **navigation.**

On a well-planned website, it's quick and easy to get to information pages — that's navigation. Plan navigation before pages are created. Establish a navigation plan to ensure that viewers quickly get what they need and that new pages of content can be quickly inserted and located.

3. Page **layout**

What's the difference between a webpage and a GREAT webpage? A webpage gives us information; a GREAT webpage catches our attention — that "Oh, Wow!" reaction — and gives information as expeditiously as possible — the key is planning and creativity.

To begin layout, analyze the information to be displayed and decide how it will be most readable. Pick the **template** that best accommodates that display. As your templates were created, page layout may have been anticipated. There are three methods to create balanced page layout: blockquote margins, tables, and **frames**. Each method has **pros and cons**; it can be advantageous to use all three to build a website.

4. **Focus on** text.

Viewers come to websites for information and if they don't get what they need, flash and glitz won't bring them back. The best websites pack essential information into well-organized and well-written text. Webpages should not, however, be too heavily written text. Surveys show that web users will not read long paragraphs of information. They prefer **concise**, bite-sized sections, clearly **delineated** so they can scan for the information they need. You should write essential content as clearly and concisely as possible with brief topic headers.

5. Use graphic images to enhance, not overpower.

Graphics are a special challenge for web designers, requiring

webpage ['webpeɪdʒ] *n.* 网页

identity [aɪ'dentəti] *n.* 图标，特性

theme [θi:m] *n.* 主题

navigation [ˌnævɪ'geɪʃn] *n.* 导航

layout ['leɪaʊt] *n.* 布置，规划

template ['templeɪt] *n.* 模板，样板

frame [freɪm] *n.* 框架

pros and cons 优缺点

focus on 注意，以……为重点

concise [kən'saɪs] *adj.* 简洁的，简明的

delineate [dɪ'lɪnieɪt] *v.* 叙述，描写

balance between overuse and **skimpiness**. A site filled with graphic images can have charm and impact. The secret for effective graphics is to stick to the theme and identity of the website.

skimpiness ['skɪmpinəs]
n. 简洁，节俭

Step Three — Putting Your New Site on the Web

1. **Domain name** registration

domain name
域名

The domain name is the address that users type into their web browsers (Internet Explorer or Netscape) to view your website. You select one or more proposed domain names such as Amazon.com or Buy.com. Your domain name must not be used by anyone else, and shorter is better. Choose a domain name that best reflects your business, products or services.

2. Publish the website.

In this step you will be given instructions for uploading your new website to a computer known as a "server" where it makes your information available to any web surfer worldwide. You will need an FTP client in order to upload files to your web server. CuteFTP (www.cuteftp.com) is highly recommended.

Skill Two

Once a host has been selected, we will "publish" your new website for accessibility to everyone on the World Wide Web.

3. **Promote** your website.

promote [prə'məʊt]
v. 宣传，推广

Website promotion involves submitting your site address and search words to the top 12 search engines, which are Netscape, Yahoo, Microsoft, Alta Vista, Ask Jeeves, AOL, Excite, Google, Goto, HotBot, Looksmart and Lycos. These engines represent over 98% of all U.S. web searches. We can create special "headers" on your website that include all the necessary search engine friendly information. Other website promotion you may establish is: placing your web address on all stationery, business cards, and all broadcast and printed advertising media. You can begin as soon as your domain name is registered.

Creating and publishing your website is just "Tip of the Iceberg". The remaining 90% is unseen below the surface. These unseen features are important to the success of your website. They include **meta tag** creation, regular search engine submission, marketing exchange programs and many others.

meta tag 中继标签

.End.

website *n.* 网站	prior *adj.* 在先的，优先的
frustration *n.* 挫折，挫败	tutorial *n.* 辅导，指导
allocate *v.* 分配	webpage *n.* 网页
identity *n.* 图标，特性	theme *n.* 主题
navigation *n.* 导航	layout *n.* 布置，规划
template *n.* 模板，样板	frame *n.* 框架
pros and cons 优缺点	focus on 注意，以……为重点
concise *adj.* 简洁的，简明的	delineate *v.* 叙述，描写
skimpiness *n.* 简洁，节俭	domain name 域名
promote *v.* 宣传，推广	meta tag 中继标签

参考译文 技能2 网站设计

许多人都希望能够创建一个吸引人的网站，但是创建一个好的网站并不是靠手指的敲击，而是要利用大脑进行设计。优秀的网站来源于全方位的设计，事先做好充分的准备既节省时间又可以避免在网页设计和更新的过程中遇到阻碍。下面的网站设计三部曲将指导你创建一个高水准的有吸引力的网站。

第一步——弄清谁使用你的网站及他们所需要的信息

好的网站都十分清楚它们的客户群是谁、为什么要访问网站，因此网站会为访问者提供一个即时的、有吸引力的界面。客户访问我们的网站不是因为我们花了时间去创建它，而是因为客户想通过访问我们的网站来获得最大的收益。

第二步——编辑你的网页

1. 创建图标并将此图标应用到所有的网页上

网站的访问者应该准确地知道我们是谁，点击相关链接后应该知道他们是否还在我们的网页上。这并不意味着每个网页看上去都一样，但是在整个网站设计中，颜色和图片应保持一致。要为你的网站创建一个主题或有标识性的特征。

2. 创建界面友好的导航

在一个好的网站里获取信息是很快、很容易的，因为那里有导航。在网页创建之前要设计好导航，建立一个导航能保证访问者快速地获取他们所需要的信息，而且新的内容也能很快地添加进去。

3. 页面布局

网页与好的网页的差别是什么呢？网页给我们提供信息；好的网页能吸引我们的注意力，使我们有“哦，哇！”的反映，并能尽可能快速地提供信息。两者的区别关键在于网页的设计与创意。

开始布局时，要分析所要显示的信息，确定如何才能使它们被更好地阅读。选择最合适的模板，一旦模板选定了，页面布局也就差不多了。这里有三种方法来创建协调的页面布局：第一种是页边缘式；第二种是表格式；第三种是框架式。每一种方法都有优缺点，最有利的方法是综合使用这三种方法来创建网站。

4. 专注文本

访问者浏览网站是为了获得信息，如果他们得不到他们所需要的信息，即使是动画或图片也不会让他们停留。好的网站能够很好地组织和撰写重要的信息。然而，网页也不能包含过多的文字。调查表明，网络用户不喜欢阅读长的段落，他们喜欢短小精悍的文章，以便能快速浏览得到所需的内容。所以应该将重要的内容尽可能写得简单明了，当然还要有一个简要的标题。

5. 利用图片增强效果，但不能过于花哨

对网站设计者来说，合理地利用图片是另外一个难点，既不能过多，又不能太少。附有图片的网站比较有吸引力和影响力。利用好图片的一个有效方法是使图片紧扣主题和网站的内容。

第三步——网站发布

1. 域名注册

域名就是用户要访问你的网站时在网络浏览器（Internet Explorer或Netscape）中输入的地址。你可以选择类似Amazon. com或Buy. com这种形式的名字作为你的域名。你的域名必须得是别人没有用过的，并且越短越好。选择最能反映你公司的业务、产品或服务的名字作为你的域名。

2. 网站发布

这步操作将指导你把新建的网站上传到服务器系统上，上传后就可以使全世界的浏览者访问到你的网站。你需要使用FTP客户端将文件上传到你的网络服务器上。CuteFTP（www. cuteftp. com）是非常值得推荐的上传工具。

主机选定后，你就可以向全世界公布你的网站了。

3. 网站的推广宣传

网站的宣传包括向12大著名的搜索引擎提交你的网址和关键字，这12个搜索引擎是Netscape、Yahoo、Microsoft、Alta Vista、Ask Jeeves、AOL、Excite、Google、Goto、HotBot、Looksmart和Lycos。这些搜索引擎代表着全美国网站搜索98%以上的份额。我们还可以为网站创建特定的标题，使其包含更方便大部分搜索引擎搜索的信息。其他的网站宣传方法有：将你的网址添加到信笺、商业名片、广播或广告媒体中。当你的域名注册完后，你就可以立即开始对网站进行推广与宣传了。

网站的创建和发布就像“冰山一角”一样。其他90%的工作都是在表面以下看不见的。对于一个成功的网站来说，这些看不见的工作也是非常重要的。这些工作包括创建中继标签，规划搜索引擎的信息提交，规范市场交易系统，以及许多其他工作。

Fast Reading One The Advancement of the Computer

Fast Reading One

The use of the transistor in computers in the late 1950s marked the coming of the second-generation computers. The most notable change was that transistors replaced vacuum tubes. This meant that the advent of smaller, faster, more reliable and less expensive computers was possible with vacuum-tube machines. In addition, the second-generation computers were given auxiliary storage, sometimes called external or secondary storage. Data was stored outside the computer on either magnetic tapes or magnetic disks. Using magnetic tapes or magnetic disks for input and output operations increased the speed of the computer.

RAM capacities increased from 8,000 to 64,000 words in commercially available machines by the 1960s, with access times of 2 to 3 ms (milliseconds). These machines were very expensive to purchase or even to rent and were particularly expensive to operate because of the cost of expanding programming. Such computers were mostly found in large computer centers operated by industry, government, and private laboratories — staffed with many programmers and support personnel.

Late in the 1960s the integrated circuit, or IC, was introduced, making it possible for many transistors to be included on one silicon chip. Therefore, the computers became even smaller and cheaper while their memory capacities became larger. The microprocessor became a reality in the mid-1970s with the large-scale integrated (LSI) circuit. The earliest microcomputer, the Altair 8800, was developed in 1975 by Ed Roberts; this machine used the Intel microprocessor and had less than 1 kilobyte of memory.

In the 1980s, very-large-scale integrated (VLSI) circuit, in which hundreds of thousands of electronic components were etched into a single silicon chip, became more and more common. Many companies, some new to the computer field, introduced in the 1970s programmable minicomputers supplied with software packages. The "shrinking" trend continued with the introduction of personal computers(PCs), some of which are programmable machines small enough and inexpensive enough to be purchased and used by individuals.

By the late 1980s, some personal computers were run by microprocessors that, handling 32 bits of data at a time, could process about 4,000,000 instructions per second. Microprocessors equipped with read-only memory(ROM), which stores constantly-used, unchanging programs, then performed an increased number of process-control, testing, monitoring, and diagnosing functions, like automobile ignition systems, automobile-engine diagnosis, and production-line inspection duties.

From the integrated circuit to large-scale integration and to very-large-scale integration, this was the start of the microprocessor age. The microprocessor continued to improve from the 80286, 80386 to the 80486, then Pentium, Pentium II and so on.

Modern digital computers are all conceptually similar, regardless of the size. They can be divided into several categories on the basis of cost and performance: the personal computer or microcomputer, a relatively low-cost machine, usually of desktop size. It also includes laptops which are small enough to fit in a briefcase and palmtops which can fit into a pocket; the workstation, a microcomputer with enhanced graphics and communications capabilities that make it especially useful for office work; the minicomputer, generally too expensive for personal use, is suitable for a business, school, or laboratory; the mainframe computer, a large, expensive machine which meets the needs of major business enterprises, government departments, scientific research establishments; the supercomputer, is the largest and fastest computer.

The "fifth-generation" computer is using new technologies in very large integration, along with new programming language, and will be capable of amazing feats in the area of artificial intelligence, voice recognition. One important parallel-processing approach is neural network, which mimics the architecture of the nervous system.

Pic 5.1 Tianhe-l

This picture (Pic 5.1) shows China's first petaflop/s scale supercomputer—Tianhe-1(TH-1). The Chinese National University of Defense Technology (NUDT) recently unveiled China's fastest supercomputer, also the world fifth fastest computer, which is able to do more than one petaflop calculation per second theoretically at its peak speed. The TH-1 is made up of 80 compute cabinets including 2,560 compute nodes and 512 operation nodes. The TH-1 system will be used to provide high performance computing service for the Tianjin area and the northeast of China. NSCC（National Supercomputer Computer Center）-TJ plans to use this system to solve the computing problems in data processing for petroleum exploration and the simulation of large aircraft designs. Other uses for the TH-1 supercomputer include the sciences, financial, automotive and shipping industries.

One continuing trend in computer development is micro-miniaturization, which is the effort to compress more circuit elements into smaller and smaller chip space.

.End.

参考译文 快速阅读1 计算机的发展

在20世纪50年代后期，晶体管的应用标志着第二代计算机的问世。其最显著的变化就是晶体管代替了电子管。使用晶体管意味着可以制造出速度更快、性能更可靠、价格更便宜的计算机，与电子管计算机相比体积也更小。另外，第二代计算机具有辅助存储器（也可称为外存或辅存）。数据可以存储在计算机外的磁带或磁盘上。使用磁带或磁盘进行输入/输出操作可以提高计算机的运算速度。

到了20世纪60年代，商用机中RAM的容量从8 000字增长到64 000字，访问时间为2～3毫秒。这些机器的价格相当昂贵，连租用费用都很高，这是因为不断扩充的程序成本使得运行费用居高不下。这样的计算机主要应用于工业、政府和私人实验室的大型计算机中心，并由许多程序员和维护人员进行操作。

20世纪60年代晚期，集成电路的引用使得许多晶体管可以嵌在一个硅片内，因此计算机的体积更小，价格更低，但存储容量却更大了。20世纪70年代中期，大规模集成电路的微处理器问世了。最早的微型计算机Altair 8800于1975年由Ed Roberts制造，这台计算机采用了英特尔微处理器，并且有不到1000字节的内存。

20世纪80年代，超大规模集成电路使成百上千的电子元件集成在一个硅片内的现象变得越来越普遍。许多公司，包括一些新涉足计算机领域的公司，都在20世纪70年代引进了由软件包支持的可编程小型计算机。随着个人计算机的产生，“收缩”的趋势仍在继续，其中一些可编程机器体积小，价格低，足以满足个人购买和使用的需求。

到了20世纪80年代后期，一些个人计算机由一次可处理32位的微处理器来运行，每秒大约可处理400万条指令。微处理器装配有只读存储器（ROM），用来存储经常使用的、不变的程序，这样的微处理器可实现越来越多的过程控制、测试、监视和诊断功能，譬如汽车点火系统、汽车引擎诊断和生产线检查等功能。

从集成电路到大规模集成电路再到超大规模集成电路，揭示着微处理器时代的开始。微处理器的型号也由80286、80386、80486一直改进到Pentium系列。

现代数字计算机无论大小，其设计理念都基本相似。依据成本和性能，基本上可以分成几类：个人计算机（微型计算机），价格不高，通常指台式机，也包括可以装入公文包的便携式计算机和可以装入衣服口袋的掌上电脑；工作站，一种具有较强的绘图和通信能力的计算机，通常在办公室使用；小型计算机，一般比较昂贵，不适合个人使用，适用于公司、学校或图书馆；大型机，是一种大型的昂贵的计算机，主要适用于企业、政府部门、科研机构等；巨型机（超级计算机），是最大型、最昂贵的计算机。

第五代计算机使用的是超大规模集成电路新技术和新程序语言，也将在人工智能、语音识别中有惊人业绩。还有一个重要的并行发展的分支是神经网络，它能够模仿人类神经网络的结构。

图5.1展示的是中国第一台每秒千万亿次超级计算机——天河一号。它是由国防科技大学（NUDT）研制出来的中国速度最快的超级计算机，在世界上排名第五，它理论上能够达到每秒

钟千万亿次的峰值速度。天河一号有80个机柜，包含2 560个计算节点和512个服务节点。天河一号将为天津以及我国的东北地区提供高性能的服务。国家超级计算机天津中心计划将天河一号应用于石油勘探、航空飞船模拟设计、科研、金融、汽车制造、运输等方面的数据处理。

计算机的一个发展趋势是小型化，即将更多的电路元件压缩在更小的芯片上。

Fast Reading Two Storage Devices

Fast Reading Two

We know that the CPU controlled by program can process data. Then where are the data and the program from? The answer is storage devices. We usually divide the storage devices into two types: the main memory and the secondary storage. A CPU can only execute the instructions of a program which has already been in the main memory.

The main memory of most computers is composed of RAM. A programmer can read and write RAM. We can store data and programs into RAM. When we have finished using them, we can let new ones occupy the position of the main memory, destroying the old ones. In a word, the content of RAM is easy to change. Sometimes we don't want the content of memory to be changed, for example, the automatic teller terminals are used in many banks. They are controlled by a small computer, which is controlled by a program. If someone can modify the data, it may give free access to certain accounts; the bank would not allow such things to happen. In fact, these programs are stored in ROM, which we can only read but cannot modify. In a word, ROM is permanent memory that can be read, but not be written. How can a program or data enter the computer system? We often use diskette drive to copy them into the main memory. Then we come to the concept of secondary storage.

Hard Disk

The hard disk is also called the hard drive, hard disk drive or fixed disk drive. The hard drive is the primary device that a computer uses to store data. Most computers have one hard drive located inside the computer case. If a computer has one hard drive, it is called "drive C". If a computer has additional hard drives, they are called "drive D, E, F", and so on. And the hard drive light is on when the computer is using the hard drive. Do not move the computer when this light is on.

The hard drive magnetically stores data on the stack of rotating disks, called platters. And a hard drive has several read/write heads that read and record data on the disks. A hard drive can store your programs and data files.

How shall we choose a hard drive? The first factor is the capacity. The amount of information a hard drive can store is measured in bytes. A hard drive with a capacity of 2GB to 20GB will suit most home and business users. Purchase the largest hard drive you can afford. A hard drive will be quickly filled up with new programs and data. For example, Microsoft Word is a word processing program that requires about 16 MB of hard drive space. The second factor is average access time. The average access time is the speed at which a hard drive finds data. It is measured in milliseconds (ms).

One millisecond equals 1/1,000 of a second. Most hard drives have an average access time of 9 to 14 ms. The less the average access time is, the faster the hard drive will be. Up to now, there are several connection types of the hand disk, such as IDE, EIDE, SCSI and so on.

Removable Hard Disk

An interesting compromise between internal and external hard disks is the removable hard disk drive tray. A tray is installed into a standard PC case drive bay that allows regular internal hard disks to be placed into it. You can then swap the internal hard disk with another one without opening up the case, allowing you to use hard disks as a removable storage medium. In a way, the concept is similar to the way a removable car stereo is designed. These trays are also commonly called mobile racks or drive caddies.

For certain applications, this is the ideal removable storage device: it uses regular hard disks, which are very fast, highly reliable, very high capacity and very inexpensive. They can be used for backup purposes.

If you decide to use a mobile rack system, be sure to check out the specifications of the unit you are considering carefully. Different models come with support for different speed drives. Some are made primarily of metal and others of plastic, and so on. Metal units will provide better cooling than plastic ones.

.End.

参考译文 快速阅读2 存储设备

我们知道CPU是在程序的控制下处理数据的。那么，数据和程序是从哪里来的呢？答案是存储设备。我们通常把存储设备划分成两类：主存储器和辅助存储器。CPU只能执行已经存放在主存中的程序指令。

大多数的计算机主存是由随机存取存储器(RAM)组成的。编辑器可读写RAM。我们可以把数据和程序存入RAM。当我们完成操作时，可以把新的内容复制到主存中。这个复制过程洗刷掉了旧内容。总之，RAM的内容容易被更改。有时我们不想改变存储器的内容。例如，用于很多银行中的自动取款机，它们由小型计算机控制，而小型计算机又由程序控制。如果有人能修改数据，便可以自由访问某些账户，银行是不允许这种事情发生的。事实上，这些程序只存放在只读存储器中，我们只能读取而不能更改。总而言之，ROM是永久性的存储器，只能读不能写。怎样能把程序和数据输入计算机系统呢？我们通常利用磁盘驱动器把它们拷贝到主存里，于是便引入了辅助存储器的概念。

硬盘

硬盘也叫作硬驱、硬盘驱动器或固定磁盘驱动器。硬盘驱动器是计算机用来存储数据的主要设备。大多数计算机都在机箱内设置一个硬盘驱动器。如果计算机只有一个硬盘驱动器，它就被称作“C盘”。如果计算机有另外的硬盘驱动器，它们则被称作“D盘、E盘、F盘”等。当计算机在使用硬盘驱动器时，硬盘指示灯是亮的。硬盘指示灯亮时不要移动计算机。

硬盘利用磁力将数据存放在旋转磁盘的轨道上，称为磁碟。硬盘上有几个用来存取数据读写的磁头。硬盘能存储程序和数据文件。

我们应该怎样选择硬盘呢？首要参数是容量。硬盘所能存储的信息数量是用字节来衡量的。2GB到20GB容量的硬盘能满足大多数的家庭和商业用户的需求。在你能负担的范围内，尽量购买大容量的硬盘。硬盘会很快被新的程序和数据占满。例如，微软Word是一个要求16MB硬盘空间的文字处理程序。第二个参数是平均存取时间，它是硬盘寻找数据的速度，以毫秒为单位。1毫秒等于1/1 000秒。大部分硬盘的平均存取时间在9～14毫秒。平均存取时间越少，硬盘的速度就越快。现有的几种硬盘连接类型有IDE、EIDE、SCSI等。

移动硬盘

在内部和外部硬盘之间折中的方法就是使用可移动硬盘驱动器托盘。托盘安装在标准的PC机箱驱动器槽上，该托盘允许一般的内部硬盘安装在上面。然后内部硬盘与另一个硬盘便可以交换却不必打开机箱，从而允许你像使用移动存储介质一样地使用硬盘。在某种程度上，它的设计类似于一套可移动的汽车立体音响。这些托盘也叫作移动架或驱动盒。

对某些应用来说，移动硬盘是理想的可移动存储设备：它使用常规硬盘，速度快，可靠性高，容量大，并且价格便宜。移动硬盘还可用于备份。

如果你决定使用一个移动硬盘系统，一定要仔细地检验你所考虑的这种硬盘的规格。不同规格的硬盘支持的速度不同，有些主要是由金属构成的，有些则是由塑料制成的，还有其他材质。金属的比塑料的更容易冷却。

Exercises

Ex 1 What do you do on the Internet? Can you give us some examples?

Ex 2 Fill in the table below by matching the corresponding Chinese or English equivalents.

protocol	
	主机
DNS	
	超文本传输协议
FTP	
	中继标签
video conferencing	
	远程登录
terminal	
	超链接

Ex 3 Choose the best answer for each of the following statements according to the text we've learnt.

1. Which protocol does the Internet mainly use? _________

    A. The OSI reference model.

    B. TCP/IP.

    C. File Transfer Protocol.

    D. HTTP.

2. An example of a client-server application on the Internet would be _________.

    A. NIC

    B. E-mail

    C. Word

    D. hard drive utilities

3. Which service dosen't the Internet provide? _________

    A. E-mail.

    B. WWW.

    C. ASP.

    D. FTP.

4. A protocol is a set of _________.

    A. drivers

    B. servers

    C. regulations

    D. hardwares

5. What is used to specify placement of texts, files, and objects that are to be transferred from web server to web browser? _________

    A. HTTP.

    B. HDLC.

    C. HTML.

    D. URL.

Part B Practical Learning

Task One

Task Two

Training Target

In this part, there are two tasks in English environment. You should complete these tasks in groups under the joint guidance of professional teachers and laboratory teachers, so as to train and improve yours ability to complete professional tasks in English environment.

Task One Collect the Necessary Data

In this task, students must collect necessary data for the online store.

The data includes the name of your store, the form of your store and the platform to open your store. Choose the product, and be clear about the details of the goods and so on.

Task Two Design the Online Store

In this task, students must design their online store. There is some information about designing online store. I hope the information can help students finish their task.

Four Things You Need to Have

1. Registered Domain
2. Premium Web Hosting Account with Cpanel and auto-installer support
3. WordPress — an open source platform you need
4. A list of your Fashion Products with their respective prices and descriptions

1. Registered Domain

To start, you need to have a professional aura, a sense of unique identity of your online store so that your potential online customers and other site visitors can locate your site. This is possible by getting a registered domain name for your site at Name.com and at any other accredited best domain registrar.

2. Premium Web Hosting Account with Cpanel and auto-installer support

Steps in Signing Up on HostGator

(1) Browse through to HostGator.com and select your web hosting plan (Pic 5.2).

Pic 5.2 select your web hosting plan

(2) Enter your domain name. Input any validated HostGator promotion code you have got (Pic 5.3).

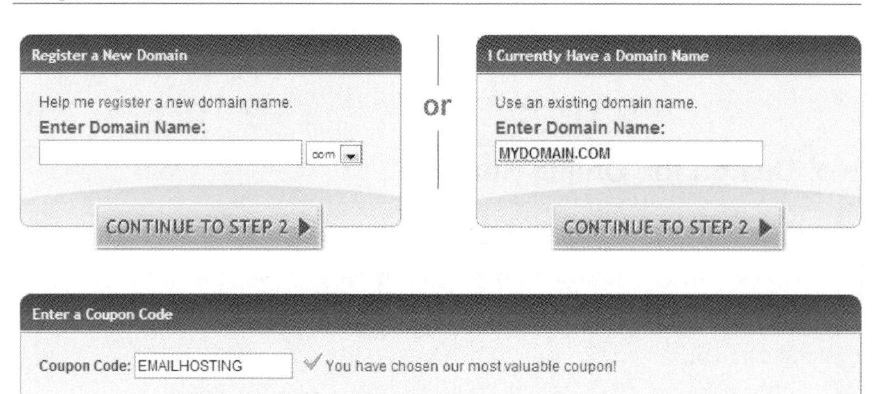

Pic 5.3 Input any validated HostGator promotion code

(3) Select Package Type and Billing Cycle (Pic 5.4).

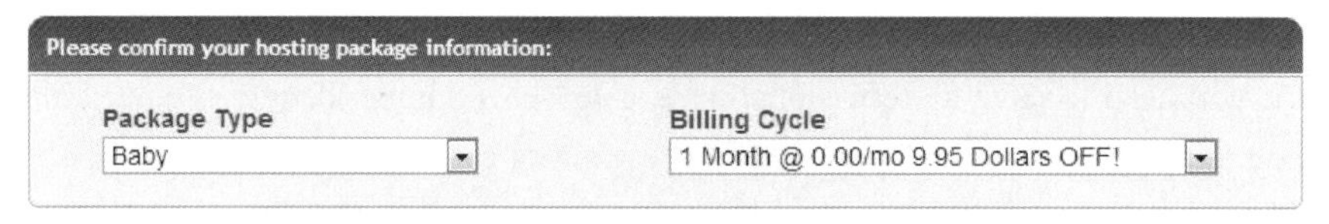

Pic 5.4 Select Package Type and Billing Cycle

(4) Make your Account along with your billing specifics (Pic 5.5).

Please choose your account information:

Username

May not contain capital letters
Must start with a letter
Must be 2-8 characters long
May not contain special characters

Security Pin

Must be 4-8 characters long
May only contain numbers

Pic 5.5 Choose your account information

(5) Finally, create your account (Pic 5.6).

Please review the order details below:

24/7/365 Phone, Live Chat, Email Support	FREE!
Instant Account Activation	FREE!
Money Back Guarantee!	45 Days
Hosting	Baby
Package (1 Month Cycle)	$9.95
Coupon Credit (EMAILHOSTING | 9.95 Dollars OFF!)	-$9.94
Hosting Addons	$0.00
Existing Domain (mydomain.com)	$0.00
Setup Price	$0.00
	Total Due: $0.01

☐ I have read and agree to the terms and conditions of use.

CREATE ACCOUNT

Pic 5.6 create your account

3. WordPress — an open source platform you need

First, install WordPress.

After installing, you can use the WordPress.

(1) Mercor (Pic 5.7) — amazing theme, you'll love the sticky memo and the excellent design.

Pic 5.7 Mercor

(2) Neighborhood (Pic 5.8) — super responsive, retina ready, and built upon the 1170px Twitter Bootstrap framework. Featuring a clean, modern, and superbly slick design, packed with the most powerful Swift Framework which offers limitless possibilities.

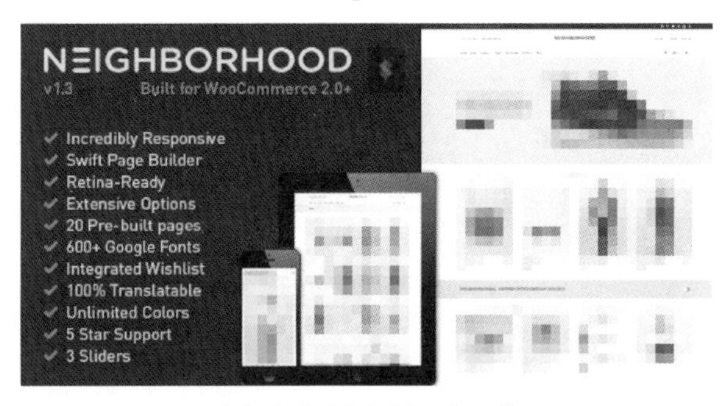

Pic 5.8 Neighborhood

(3) Primashop (Pic 5.9)—future-proof WooCommerce compatible and it is compatible with most of WooCommerce extensions.

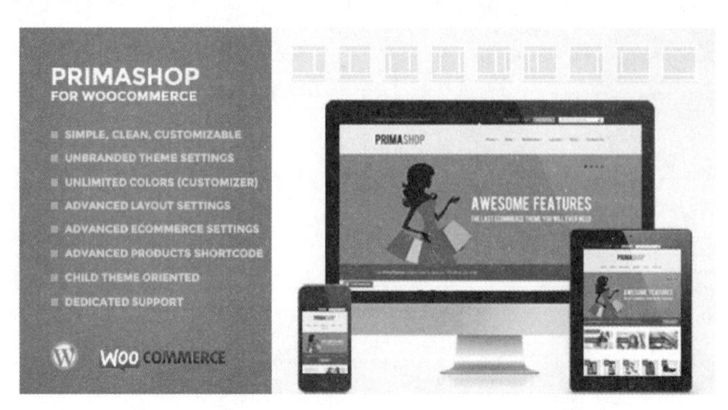

Pic 5.9 Primashop

(4) Camp (Pic 5.10)—easy-to-customize and fully featured E-Commerce WordPress Theme.

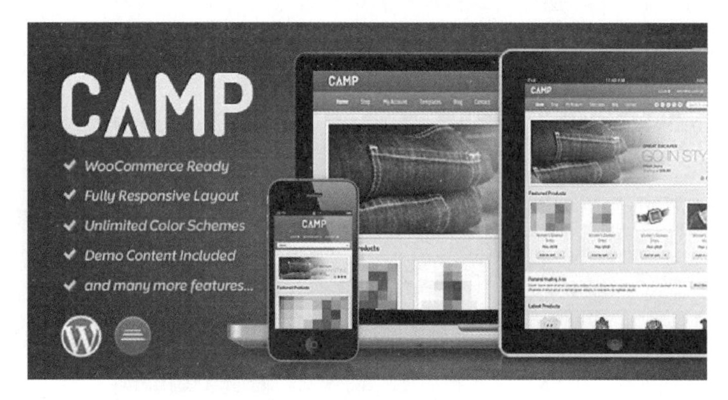

Pic 5.10 Camp

4. List of your Fashion Products with their respective prices and descriptions

After you install everything you need and customize what needs to be customized, it's time for you to add your products. Your listing becomes handy by adding items on your site. Most E-Commerce will add a menu tab for the items. So simply head into that menu, then place your product title, price, description and most importantly your product image. You have to do manually the adding of your own products. Add and edit your products as often as you need to.

Congratulations, you have successfully created your own online store!

Part C Occupation English

Occupation English

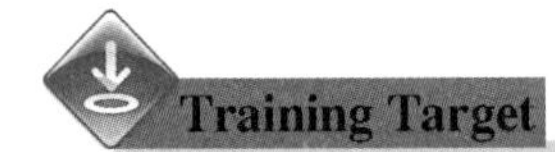

In this part, there is an English dialogue in real life and work environment. You will play the roles of A, B and C and read the dialogue aloud to practice your ability to use English.

How to Design a Website? 如何制作一个网站?

Role Setting: Website Producer (A)，Website learner（B）and Website Designer（C)

角色设置：电脑培训学校服务人员（A），要学习网站设计人员（B）和网站制作人员（C）

作为网络客服人员，需要向顾客提供各种有关上网的咨询服务。网络连接是常见的问题之一。

A: Hello, welcome. What can I do for you? 您好，欢迎光临，我能帮你什么？

B: Hello, I would like to ask about the website design. 您好，我想咨询下网站设计方面的问题。

A: No problem. There are teachers in charge of website design in our school. I'll take you there. Please follow me. It's the second office on the left of the stairs. 没问题，我们学校有专门负责网站设计方面的老师，我带您过去。请跟我来，在楼梯左边第二个办公室。

B: OK 好的。

A: Hello, Mr. Wang, this gentleman wants to learn about website design. Can you give him an answer?

C: Oh, welcome. Come in, please . That's what I should do. 哦，欢迎。请进。那是我应该做的。

C: Please have a seat, sir. Would you like to learn website design? 先生您请坐，您想学习网站设计？

B: Yes. 是的。

C: First of all, learning to make a website must have a certain computer foundation, such as being able to operate computer skillfully, edit words and so on. 首先，学习制作网站要有一定的电脑基础，比如能熟练操作电脑，会文字编辑等。

B: Yes, there is no problem with the basic operation of the computer. You can rest assured. 是的，对于电脑基本操作这方面完全没有问题的，这个您放心。

C: That is much easier to do. Specialized learning website design is good. First of all, I'll introduce to you what you have to learn, and then you can consider whether you can learn it. 那就好办多了，专门学习网站设计就好了。我先给你介绍下都得学习那些内容，您再考虑能不能学。

B: OK.好的

C: The first stage: at the beginning, it is best to learn some web editing software and basic web script syntax, such as Dreamweaver software, HTML syntax (Baidu: HTML introduction), CSS syntax (Baidu: CSS introduction), so that you can better understand the principle of web page production and operation. 第一阶段:开始时最好是学些网页编辑软件和基础网页脚本语法,如Dreamweaver软件,HTML语法(百度一下:HTML入门),CSS语法(百度一下:CSS入门),这样可以使您更了解网页制作与运营的原理。

The second stage: after learning the web page making software and basic grammar, in order to make the website more beautiful and powerful, you also need to learn some dynamic programming languages; commonly used are ASP, PHP and ASP.NET Etc. 第二阶段:学会了网页制作软件和基础语法后,为了让网站更漂亮、功能更强大,你还需要学一些动态编程语言,常用的有ASP, PHP和ASP.NET等。

The third stage: combine your own knowledge to develop the website. 第三阶段:结合你自己已掌握的知识来开发网站了。

C: Through my introduction, do you know what you want to learn? 通过我的介绍,您对要学习的内容了解了吗?

B: It's very clear. I have also searched many related problems on the Internet. Thank you for your introduction. I'll take a look at the basic programming, and then I'll learn. It'll be easier and faster. 非常清楚了,我自己也在网上查找了不少相关方面的问题。那谢谢您的介绍,我先看下基本编程方面的,然后再来学习,这样能快点、容易点。

C: All right. You are welcome at any time. 好的,随时欢迎您的光临。

Word Building

前缀/后缀由一个或几个字母组成,放在词根或单词之前/之后,组成一个新词。

(1) inter-(前缀): 在……之间;中间

connect 联系 ———— interconnect 互联

act 行动,起作用 ———— interact 互相作用

(2) hyper-(前缀): 超出

text 文本 ———— hypertext 超文本

link 链接 ———— hyperlink 超链接

(3) un-(前缀): 反、不、非

format 格式化 ———— unformat 未格式化

delete 删除 ———— undelete 反删除,恢复删除

(4) -ness(后缀): 情况,性质

ill 有病的 ———— illness 疾病

idle 懒惰的 ———— idleness 惰性

Ex Try your best to guess the meaning of each word on the right according to the clues given on the left.

national	国家的（形容词）	international	________
media	媒体（名词）	hypermedia	________
install	安装（动词）	uninstall	________
repository	仓库（名词）	repost	________
unprecedented	前所未有的（形容词）	precedence	________
retailer	零售商（名词）	retail	________
notification	通知（名词）	notify	________
norm	标准（名词）	normal	________
skimp	节俭的（形容词）	skimpiness	________
graphic	图形的（形容词）	graph	________

Exercises

Ex 1 You have tried to create your own website. But once you have owned your website, how to make it perfect?

Ex 2 Fill in the table below by matching the corresponding Chinese or English equivalents.

layout	
	网站
homepage	
	网页
domain name	
	框架
navigation	
	模板
web address	

Ex 3 Choose the best answer for each of the following statements according to the text we've learnt.

1. What does the word "frustration" mean in the first paragraph of *Website Design*? ________
   A. Lacking of programming knowledge
   B. Setting up a website is very difficult
   C. Not having extensive planning
   D. Having no time to surf the Internet

2. Which of the following statements is not the step of website design? _________

   A. Determining consumers

   B. Webpage design

   C. Surfing the World Wide Web

   D. Publishing the website

3. Which of the following statements is not the step that you will follow when editing your webpage? _________

   A. Creating user-friendly navigation

   B. Page layout

   C. Publishing the website

   D. Using graphics

4. When you select domain name for your website, what is the important idea? _________

   A. Not be used by anyone else

   B. Best reflect your business or products

   C. Not too long

   D. All of the above

5. Which of the following is not the search engine? _________

   A. Google

   B. Baidu

   C. Netscape

   D. Dreamweaver

Project Six

Creating Database

Part A Theoretical Learning

Part B Practical Learning

Part C Occupation English

Part A Theoretical Learning

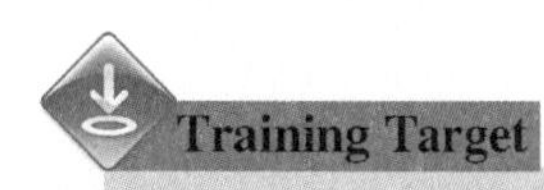

In this part, our target is to improve the speed of reading professional articles and the comprehension ability of the reader. We have marked specialized key words and some flexible sentences. Try to grasp the main idea of each paragraph.

Skill One Foundation of Database

A **database** is an organized collection of **data**. It is the **collection** of **schemas**, tables, queries, reports, views, and other objects. **The data is typically organized to model aspects of reality in a way that supports processes requiring information, such as modelling the availability of rooms in hotels in a way that supports finding a hotel with vacancies.**

Formally, a "database" refers to a set of related data and the way it is organized. Access to this data is usually provided by a "**database management system**" (**DBMS**) consisting of an integrated set of computer software that allows users to interact with one or more databases and provides access to all of the data contained in the database (although restrictions may exist that limit access to particular data). The DBMS provides various functions that allow entry, storage and retrieval of large **quantities** of information and provides ways to manage how that information is organized.

Because of the close relationship between them, the term "database" is often used casually to refer to both a database and the DBMS used to manipulate it.

DBMS:

A database management system (DBMS) is a computer software application that interacts with the user, other applications, and the database itself to capture and analyze data. A general-purpose DBMS is designed to allow the definition, creation, querying, update, and administration of databases. Well-known DBMSs include MySQL, PostgreSQL, MongoDB, MariaDB, Microsoft SQL Server, Oracle, Sybase, SAP HANA, MemSQL,

data ['deɪtə] *n.* 数据
database ['deɪtəbeɪs] *n.* 数据库
collection [kə'lekʃn] *n.* 收集，集合
schema ['ski:mə] *n.* 提要
vacancy ['veɪkənsi] *n.* 空房，空间
database management system *n.* 数据库管理系统
entry ['entri] *n.* 进入
retrieval [rɪ'tri:vl] *n.* 数据检索
quantity ['kwɒntəti] *n.* 大量

Skill One

SQLite and IBM DB2. A database is not generally portable across different DBMSs, but different DBMS can interoperate by using standards such as SQL and ODBC or JDBC to allow a single application to work with more than one DBMS. Database management systems are often classified according to the database model that they support; the most popular database systems since the 1980s have all supported the relational model as represented by the SQL language. Sometimes a DBMS is loosely referred to as a "database".

portable ['pɔːtəbl] *adj.* 便携式的；手提的
n. 便携机
interoperate [intə'ɒpəreɪt]
v. 互操作；互通

Existing DBMSs provide various functions that allow management of a database and its data which can be classified into four main functional groups:

•Data – Creation, modification and removal of definitions that define the organization of the data.

•Update – Insertion, modification, and deletion of the actual data.

•Retrieval – Providing information in a form directly usable or for further processing by other applications. The retrieved data may be made available in a form basically the same as it is stored in the database or in a new form obtained by altering or combining existing data from the database.

•Administration – Registering and monitoring users, enforcing data security, monitoring performance, maintaining data integrity, dealing with concurrency control, and recovering information that has been corrupted by some event such as an unexpected system failure.

removal [rɪ'muːvl]
n. 移动；调动
update [ˌʌp'deɪt]
v. 使现代化；更新
insertion [ɪn'sɜːʃn]
n. 插入排序
monitor ['mɒnɪtə(r)]
v. 监视
maintain [meɪn'teɪn]
v. 维持；保持
integrity [ɪn'tegrəti]
n. 完整；完好
concurrency [kən'kʌrənsi]
n. 并发性

Database model:

A database model is a type data model that determines the logical structure of a database and fundamentally determines in which manner data can be stored, organized,and manipulated. The most popular example of a database model is the relational model (or the approximation of relational SQL), which uses a table-based format.

fundamentally [ˌfʌndə'mentəli] *adv.* 根本上；完全地

Database language:

Database languages are special-purpose languages, which do one or more of the following:

Data definition language – defines data types such as creating, altering, or dropping and the relationships among them.

Data manipulation language – performs tasks such as inserting, updating, or deleting data occurrences.

Query language – allows searching for information and computing derived information.

Database languages are specific to a particular data model.Notable examples include:

alter ['ɔːltə(r)] *v.* （使）改变，更改
drop ['drɒp] *v.* 落下
occurrence [ə'kʌrəns]
n. 发生；出现；存在
derive [dɪ'raɪv]
v. 获得；取得

SQL combines the roles of data definition, data manipulation, and query in a single language. It was one of the first **commercial** languages for the relational model, although it departs in some respects from the relational model as described by Codd (for example, the rows and columns of a table can be ordered). SQL became a standard of the American National Standards Institute (ANSI) in 1986, and of the International Organizationfor Standardization (ISO) in 1987. The standards have been regularly enhanced since and is supported (with varying degrees of conformance) by all mainstream commercial relational DBMSs.

commercial [kə'mɜːʃl] *adj.* 贸易的；商业的

OQL is an object model language standard. It has influenced the design of some of the newer query languages like JDOQL and EJB QL.

XQuery is a standard XML query language **implemented** by XML database Systems such as MarkLogic and Exist, by relational databases with XML capability such as Oracle and DB2, and also by in-memory XML processors such as Saxon.

implement ['ɪmplɪment] *v.* 执行

SQL/XML combines XQuery with SQL.

A database language may also **incorporate** features like:

DBMS-specific Configuration and storage engine management **computations** to modify query results, like counting, summing, averaging, sorting grouping, and cross-**referencing**.

Constraint enforcement (e.g. in an automotive database, only allowing one engine type per car).

Application programming interface version of the query language, for programmer convenience.

incorporate [ɪn'kɔːpəreɪt] *v.* 将……包括在内；包含

computation [ˌkɒmpju'teɪʃn] *n.* 计算

reference ['refrəns] *v.* 查阅；参考

constraint [kən'streɪnt] *n.* 约束；约束条件

enforcement[ɪn'fɔːsmənt] *n.* 执行；实施

.End.

data *n.* 数据　　database *n.* 数据库
collection *n.* 收集，集合　　schema *n.* 提要
vacancy *n.* 空房，空间　　entry *n.* 进入
database management system *n.* 数据库管理系统
retrieval *n.* 数据检索　　quantity *n.* 大量
portable *adj.* 便携式的，手提的 *n.* 便携机
interoperate *v.* 互操作；互通　　removal *n.* 移动；调动

Key Words

update *v.* 使现代化，更新
monitor *v.* 监视
integrity *n.* 完整；完好
fundamentally *adv.* 根本上；完全地
drop *v.* 落下
derive *v.* 获得；取得
implement *v.* 执行
computation *n.* 计算
constraint *n.* 约束；约束条件
insertion *n.* 插入排序
maintain *v.* 维持；保持
concurrency *n.* 并发性
alter *v.* （使）改变，更改
occurrence *n.* 发生；出现；存在
commercial *adj.* 贸易的；商业的
incorporate *v.* 将……包括在内；包含
reference *v.* 查阅；参考
enforcement *n.* 执行；实施

参考译文 技能1 数据库基础

数据库是有组织的数据集合。它是模式、表、查询、报表、视图和其他对象的集合。这些数据通常被组织起来，以一种支持需要信息的过程的方式来模拟现实的各个方面，例如以一种支持查找有空房的酒店的方式来模拟酒店的客房可用性。

在形式上，“数据库”指的是一组相关数据及其组织方式。访问这些数据通常由“数据库管理系统”（DBMS）提供，该系统由一套集成的计算机软件组成，允许用户与一个或多个数据库交互，并提供对数据库中包含的所有数据的访问（尽管可能存在限制访问特定数据的限制）。数据库管理系统提供各种功能，允许输入、存储和检索大量信息，并提供管理信息组织方式的方法。

由于它们之间的密切关系，“数据库”一词经常被随意地用来指代数据库和用来操作它的数据库管理系统。

数据库管理系统：

数据库管理系统（DBMS）是一种计算机软件应用程序，它与用户、其他应用程序以及数据库本身交互，以捕获和分析数据。一个通用的数据库管理系统被设计成允许定义、创建、查询、更新和管理数据库。著名的数据库管理系统包括MySQL、PostgreSQL、MongoDB、MariaDB、Microsoft SQL Server、Oracle、Sybase、SAP HANA、MemSQL、SQLite和IBM DB2。数据库通常不能跨不同的DBMS移植，但是不同的DBMS可以通过使用SQL和ODBC或JDBC等标准进行互操作，以允许单个应用程序与多个DBMS协同工作。数据库管理系统通常根据它们所支持的数据库模型进行分类；自20世纪80年代以来最流行的数据库系统都支持SQL语言表示的关系模型。有时DBMS被松散地称为“数据库”。

现有数据库管理系统提供各种功能，允许管理数据库及其数据，这些功能可分为四个主要功能组：

数据：创建、修改和删除定义数据组织的定义。

更新：插入、修改和删除实际数据。

检索：以可直接使用的形式提供信息或供其他应用程序进一步处理。检索到的数据可以基本上与它存储在数据库中的形式相同，也可通过改变或合并数据库的现有数据而获得新形式。

管理：注册和监视用户，加强数据安全性，监视性能，维护数据完整性，处理并发控制，并恢复因意外系统故障等事件而损坏的信息。

数据库模型：

数据库模型是一种类型数据模型，它决定了数据库的逻辑结构，并从根本上决定了数据的存储、组织和操作方式。关系数据库模型是最流行的一个例子（或与关系模型SQL近似），它使用基于表的格式。

数据库语言：

数据库语言是一种特殊用途的语言，可以执行以下一种或多种操作：

数据定义语言——定义数据类型，如创建、更改或删除它们之间的关系。

数据操作语言——执行诸如插入、更新或删除数据的任务。

查询语言——允许搜索信息和计算派生信息。

数据库语言是专用于特定数据的模型。包括：

SQL将数据定义、数据操作和查询的角色组合在一个语言。它是关系模型最早的商业语言之一，尽管在某些方面偏离了由Codd描述的关系模型（例如表的行和列可以排序）。1986年，SQL成为美国国家标准协会（ANSI）的标准，1987年成为国际标准化组织（ISO）的标准。美国人的标准美国国家标准协会（ANSI）和国际组织，1987年标准化组织（ISO）。自此以后，这些标准得到了定期的提高并且得到了所有主流商业关系数据库管理系统的支持（不同程度的一致性）。

OQL是一种对象模型语言标准。它影响了一些较新的查询语言，如JDOQL和EJBQL的设计。

XQuery是一种标准的XML查询语言，由MarkLogic和Exist等XML数据库系统、具有XML功能的关系数据库（如Oracle和DB2）以及内存中的XML处理器（如Saxon）实现。

SQL/XML结合了XQuery和SQL。

数据库语言还可以包含以下功能：

特定于DBMS的配置和存储引擎管理。

用于修改查询结果的计算，如计数、求和、平均、排序、分组和交叉引用。

强制结束（例如，在汽车数据库中，每辆车只允许一种发动机类型）。

应用程序编程接口的查询语言版本，为程序员提供方便。

Skill Two Database model

A database model is a type of data model that determines the logical structure of a database and fundamentally determines in which manner data can be stored, organized, and manipulated. The most popular example of a database model is the relational model (or the approximation of relational SQL), which uses a table-based format.

model ['mɒdl] *n.* 模型

structure ['strʌktʃə(r)] *n.* 结构

Common logical data models for databases include:

Navigational databases

A navigational database is a type of database in which records or objects are found primarily by following references from other objects. They were a common type of database in the **era** when data was stored on **magnetic** tape; the navigational references told the computer where the next record on the tape was stored, allowing fast-forwarding (and in some cases, reversing) through the records without having to read every record along the way to see if it matched a given **criterion**.

The introduction of low-cost hard drives that provided semi-random access to data led to new models of database storage better suited to these devices. Among these, the relational database and especially SQL became the **canonical** solution from the 1980s through to about 2010. At that time a reappraisal of the entire database market began, the various NoSQL concepts, which have led to the navigational model being reexamined. **Offshoots** of the concept, especially the graph database, are finding new uses in modern transaction processing **workloads**.

Relational model database

The relational model (RM) for database management is an approach to managing data using a structure and language **consistent** with first-order **predicate** logic, first described in 1969 by Edgar F. Codd, where all data is represented in terms of tuples, grouped into relations. A database organized in terms of the relational model is a relational database.

In the relational model, related records are linked together with a "key".

The purpose of the relational model is to provide a **declarative** method for **specifying** data and queries: users directly state what information the database contains and what information they want from it, and let the database management system software take care of describing data structures for storing the data and retrieval procedures for answering queries.

Most relational databases use the SQL data definition and query language; these systems **implement** what can be regarded as an engineering **approximation** to the relational model. A table in an SQL database schema corresponds to a predicate variable; the contents of a table to a relation; key constraints, other constraints, and SQL queries

navigational [ˌnævɪ'geɪʃənl] *adj.* 导航的

era ['ɪərə] *n.* 时代

magnetic [mæg'netɪk] *adj.* 像磁铁的，有磁性的

criterion [kraɪ'tɪəriən] *n.* 标准

canonical [kə'nɒnɪkl] *adj.* 经典

offshoot ['ɒfʃu:t] *n.* 分支

workload ['wɜ:kləʊd] *n.* 工作量

consistent [kən'sɪstənt] *adj.* 一致的

predicate ['predɪkət] *n.* 谓语

declarative [dɪ'klærətɪv] *adj.* 陈述的

specify ['spesɪfaɪ] *v.* 具体说明详述；详列

implement ['ɪmplɪment] *v.* 使生效

approximation [əˌprɒksɪ'meɪʃn] *n.* 近似值

correspond to predicates. However, SQL databases deviate from the relational model in many details, and Codd fiercely argued against **deviations** that compromise the original **principles**.

correspond [ˌkɒrəˈspɒnd] v. 相一致;符合
deviation [ˌdiːviˈeɪʃn] n. 背离;偏离
principle [ˈprɪnsəpl] n. 原则，规范

The relational model was the first database model to be described in formal mathematical terms. Hierarchical and network databases existed before relational databases, but their specifications were relatively informal. After the relational model was defined, there were many attempts to compare and contrast the different models, and this led to the emergence of more rigorous descriptions of the earlier models; though the procedural nature of the data manipulation interfaces for hierarchical and network databases limited the scope for formalization.

Skill Two

Object model database

An object database (also object-oriented database management system, OODBMS) is a database management system in which information is represented in the form of objects as used in object-**oriented** programming. Object databases are different from relational databases which are table-oriented. Object-relational databases are a hybrid of both approaches.

orient [ˈɔːriənt] v. 朝向;面对

Object databases have been considered since the early 1980s.

Star schema database

In computing, the star schema is the simplest style of data mart schema and is the approach most widely used to develop data **warehouses** and **dimensional** data marts. The star schema consists of one or more fact tables referencing any number of dimension tables. The star schema is an important special case of the **snowflake** schema, and is more effective for handling simpler queries.

warehouse [ˈweəhaʊs] n. 仓库
dimensional [daɪˈmenʃənl] adj. 空间的；尺寸的
snowflake [ˈsnəʊfleɪk] n. 雪花

The star schema gets its name from the physical model's resemblance to a star shape with a fact table at its center and the dimension tables surrounding it representing the star's points.

In addition to the above main database models, there are several other types of database models, such as Associative model, Multidimensional model, XML database and so on.

.End.

Key Words

model *n.* 模型	structure *n.* 结构
navigational *adj.* 导航的	era *n.* 时代
magnetic *adj.* 像磁铁的，有磁性的	criterion *n.* 标准

canonical *adj.* 经典的	offshoot *n.* 分支
workload *n.* 工作量	consistent *adj.* 一致的
predicate *n.* 谓语	declarative *adj.* 陈述的
specify *v.* 具体说明详述；详列	implement *v.* 使生效
approximation *n.* 近似值	correspond *v.* 相一致；符合
deviation *n.* 背离；偏离	principle *n.* 原则，规范
orient *v.* 朝向；面对	warehouse *n.* 仓库
dimensional *adj.* 空间的；尺寸的	snowflake *n.* 雪花

参考译文 技能2 数据库模型

数据库模型是一种数据模型，它决定了数据库的逻辑结构，并从根本上决定了可以以何种方式存储、组织和操作数据。数据库模型最流行的例子是关系模型（或关系模型的SQL近似），它使用基于表的格式。

常见的逻辑数据模型的数据库包括：

导航数据库

导航数据库是一种数据库类型，其中记录或对象主要是通过跟踪其他对象的引用来找到的。在数据存储在磁带上的那个时代，它们是一种常见的数据库类型；导航参考告诉计算机磁带上下一条记录的存储位置，允许快速转发（在某些情况下，可以反转）记录，而不必一路读取每一条记录，看它是否符合给定的标准。

低成本硬盘的引入提供了对数据的半随机访问，导致了更适合这些设备的新的数据库存储模型。其中，关系数据库，尤其是SQL成为从20世纪80年代到2010年的标准解决方案。当时，对整个数据库市场的重新评估开始了，各种非关系型数据库概念导致导航模型被重新审视。这个概念的分支，特别是图形数据库，正在现代事务处理工作负载中找到新的用途。

关系型数据库

数据库管理的关系模型（RM）是一种使用与一阶谓词逻辑一致的结构和语言管理数据的方法，该方法在1969年由Edgar F. Codd首次描述，其中所有数据都以元组的形式表示，并分组到关系中。按照关系模型组织的数据库是关系数据库。

在关系模型中，相关记录用一个“键”链接在一起。

关系模型的目的是为指定数据和查询提供一种声明性方法：用户直接声明数据库包含哪些信息以及他们希望从中获取哪些信息，数据库管理系统软件负责描述存储数据的数据结构和回答查询的检索过程。

大多数关系数据库使用SQL数据定义和查询语言；这些系统实现了可以看作是关系模型的

工程近似。SQL数据库模式中的表对应于谓词变量；表的内容对应于关系；键约束、其他约束和SQL查询对应于谓词。然而，SQL数据库在许多细节上都偏离了关系模型，Codd强烈反对那些破坏原有原则的偏差。

关系模型是第一个用正式数学术语描述的数据库模型。层次数据库和网络数据库在关系数据库之前就已经存在，但它们的规范相对非正式。在定义关系模型之后，人们曾多次尝试对不同的模型进行比较和对比，这导致了对早期模型更严格的描述；尽管层次和网络数据库的数据操作接口的过程性限制了形式化的范围。

对象模型数据库

对象数据库（又称面向对象数据库管理系统，OODBMS）是一种数据库管理系统，其中的信息以面向对象编程中使用的对象的形式表示。对象数据库不同于面向表的关系数据库。对象关系数据库是这两种方法的混合体。

自20世纪80年代初以来，人们就开始考虑对象数据库。

星型模式数据库

在计算中，星型模式是数据集市模式中最简单的样式，也是开发数据仓库和维度数据集市最广泛使用的方法。星型模式由一个或多个引用任意数量维度表的事实表组成。星型模式是雪花模式的一个重要特例，对于处理更简单的查询更有效。

星型模式的名字来源于物理模型的相似性，它的中心是一个事实表，周围的维度表代表恒星的点。

除了以上几种主要的数据库模型外，还有其他几种类型的数据库模型，如关联模型、多维模型、XML数据库模型等。

Fast Reading One | Database Management System

Fast Reading One

A collection of interrelated data together with a set of programs to access the data, also called database system, or simply database. The primary goal of such a system is to provide an environment that is both convenient and efficient to use in retrieving and storing information.

A database management system (DBMS) is designed to manage a large body of information. Data management involves both defining structures for storing information and providing mechanisms for manipulating the information. In addition, the database system must provide for the safety of the stored information, despite system crashes or attempts at unauthorized access. If data is to be shared among several users, the system must avoid possible anomalous results due to multiple users concurrently accessing the same data.

Examples of the use of database systems include airline reservation systems, company payroll and employee information systems, banking systems, credit card processing systems, and sales and order tracking systems.

A major purpose of a database system is to provide users with an abstract view of the data. That is, the system hides certain details of how the data is stored and maintained. Thereby, data can be stored in complex data structures that permit efficient retrieval, yet users see a simplified and easy-to-use view of the data. The lowest level of abstraction, the physical level, describes how the data is actually stored and details the data structures. The next-higher level of abstraction, the logical level, describes what data is stored, and what relationships exist among those data. The highest level of abstraction, the view level, describes parts of the database that are relevant to each user; application programs used to access a database form part of the view level.

The overall structure of the database is called the database schema. The schema specifies data, data relationships, data semantics, and consistency constraints on the data.

Underlying the structure of a database is the logical data model: a collection of conceptual tools for describing the schema.

The entity-relationship data model is based on a collection of basic objects, called entities, and of relationships among these objects. An entity is a "thing" or "object" in the real world that is distinguishable from other objects. For example, each person is an entity, and bank accounts can be considered entities. Entities are described in a database by a set of attributes. For example, the attributes account-number and balance describe one particular account in a bank. A relationship is an association among several entities. For example, a depositor relationship associates a customer with each of her accounts. The set of all entities of the same type and the set of all relationships of the same type are termed an entity set and a relationship set, respectively.

Like the entity-relationship model, the object-oriented model is based on a collection of objects. An object contains values stored in instance variables within the object. An object also contains bodies of code that operate on the object. These bodies of code are called methods. The only way in which one object can access the data of another object is by invoking a method of that other object. This action is called sending a message to the object. Thus, the call interface of the methods of an object defines that object's externally visible part. The internal part of the object—the instance variables and method code—are not visible externally. The result is two levels of data abstraction, which are important to abstract away (hide) internal details of objects. Object-oriented data models also provide object references which can be used to identify (refer to) objects.

In record-based models, the database is structured in fixed-format records of several types. Each record has a fixed set of fields. The three most widely accepted record-based data models are the relational, network, and hierarchical models. The latter two were widely used once, but are of declining importance. The relational model is very widely used. Databases based on the relational model are called relational databases.

The relational model uses a collection of tables (called relations) to represent both data and the relationships among those data. Each table has multiple columns, and each column has a unique name.

Each row of the table is called a tuple, and each column represents the value of an attribute of the tuple.

The size of a database can vary widely, from a few megabytes for personal databases, to gigabytes (a gigabyte is 1 000 megabytes) or even terabytes (a terabyte is 1 000 gigabytes) for large corporate databases.

The information in a database is stored on a nonvolatile medium that can accommodate large amounts of data; the most commonly used such media are magnetic disks. Magnetic disks can store significantly larger amounts of data than main memory, at much lower costs per unit of data.

To improve reliability in mission-critical systems, disks can be organized into structures generically called redundant arrays of independent disks (RAID). In a RAID system, data is organized with some amount of redundancy (such as replication) across several disks. Even if one of the disks in the RAID system were to be damaged and lose data, the lost data can be reconstructed from the other disks in the RAID system. See Computer Storage Technology.

Logically, data in a relational database are organized as a set of relations, each relation consisting of a set of records. This is the view given to database users. The underlying implementation on disk (hidden from the user) consists of a set of files. Each file consists of a set of fixed-size pieces of disk storage, called blocks. Records of a relation are stored within blocks. Each relation is associated with one or more files. Generally a file contains records from only one relation, but organizations where a file contains records from more than one relation are also used for performance reasons.

One way to retrieve a desired record in a relational database is to perform a scan on the corresponding relation; a scan fetches all the records from the relation, one at a time.

Accessing desired records from a large relation using a scan on the relation can be very expensive. Indices are data structures that permit more efficient access of records. An index is built on one or more attributes of a relation; such attributes constitute the search key. Given a value for each of the search-key attributes, the index structure can be used to retrieve records with the specified search-key values quickly. Indices may also support other operations, such as fetching all records whose search-key values fall in a specified range of values.

A database schema is specified by a set of definitions expressed by a data-definition language. The result of execution of data-definition language statements is a set of information stored in a special file called a data dictionary. The data dictionary contains metadata, that is, data about data. This file is consulted before actual data is read or modified in the database system. The data-definition language is also used to specify storage structures and access methods.

Data manipulation is the retrieval, insertion, deletion, and modification of information stored in the database. A data-manipulation language enables users to access or manipulate data as organized by the appropriate data model. There are basically two types of data-manipulation languages: Procedural data-manipulation languages require a user to specify what data is needed and how to get those data; nonprocedural data-manipulation languages require a user to specify what data is needed without specifying how to get those data.

A query is a statement requesting the retrieval of information. The portion of a data-manipulation language that involves information retrieval is called a query language. Although technically incorrect, it is common practice to use the terms query language and data-manipulation language synonymously.

Database languages support both data-definition and data-manipulation functions. Although many database languages have been proposed and implemented, SQL has become a standard language supported by most relational database systems. Databases based on the object-oriented model also support declarative query languages that are similar to SQL.

SQL provides a complete data-definition language, including the ability to create relations with specified attribute types, and the ability to define integrity constraints on the data.

Query By Example (QBE) is a graphical language for specifying queries. It is widely used in personal database systems, since it is much simpler than SQL for nonexpert users.

Forms interfaces present a screen view that looks like a form, with fields to be filled in by users. Some of the fields may be filled automatically by the forms system. Report writers permit report formats to be defined, along with queries to fetch data from the database; the results of the queries are shown formatted in the report. These tools in effect provide a new language for building database interfaces and are often referred to as fourth-generation languages (4GLs). See Human-Computer Interaction.

Often, several operations on the database form a single logical unit of work, called a transaction. An example of a transaction is the transfer of funds from one account to another. Transactions in databases mirror the corresponding transactions in the commercial world.

Traditionally database systems have been designed to support commercial data, consisting mainly of structured alphanumeric data. In recent years, database systems have added support for a number of nontraditional data types such as text documents, images, and maps and other spatial data. The goal is to make databases universal servers, which can store all types of data. Rather than add support for all such data types into the core database, vendors offer add-on packages that integrate with the database to provide such functionality.

.End.

参考译文 快速阅读1 数据库管理系统

一组相互关联的数据和一组访问数据的程序，也称为数据库系统，或简称数据库。这种系统的主要目标是提供一个既方便又高效的环境来检索和存储信息。

数据库管理系统（DBMS）是为管理大量信息而设计的。数据管理包括定义存储信息的结构和提供操作信息的机制。此外，数据库系统必须提供存储信息的安全性，尽管系统崩溃或试图进行未经授权的访问。如果要在多个用户之间共享数据，系统必须避免由于多个用户同时访问同一数据而可能出现的异常结果。

使用数据库系统的例子包括航空公司预订系统、公司工资单和雇员信息系统、银行系统、信用卡处理系统以及销售和订单跟踪系统。

数据库系统的主要目的是为用户提供数据的抽象视图。也就是说，系统隐藏了如何存储和维护数据的某些细节。因此，数据可以存储在允许有效检索的复杂数据结构中，但用户看到的是一个简化和易于使用的数据视图。抽象的最低层次，即物理层，描述数据实际是如何存储的，并详细说明数据结构。下一个更高的抽象级别，即逻辑级别，描述存储哪些数据，以及这些数据之间存在哪些关系。最高的抽象级别，视图级别，描述了数据库中与每个用户相关的部分，用于访问数据库的应用程序构成视图层的一部分。

数据库的整体结构称为数据库模式。模式指定数据、数据关系、数据语义和数据的一致性约束。

数据库结构的底层是逻辑数据模型：用于描述模式的概念性工具的集合。

实体关系数据模型基于基本对象（称为实体）和这些对象之间关系的集合。实体是现实世界中与其他物体不同的“事物”或“对象”。例如，每个人都是一个实体，银行账户可以被视为实体。实体在数据库中由一组属性描述。例如，属性account-number和balance描述银行中的一个特定账户。关系是多个实体之间的关联。例如，储户关系将客户与她的每个账户关联起来。同一类型的所有实体的集合和同一类型的所有关系的集合分别称为实体集和关系集。

与实体关系模型一样，面向对象模型是基于对象集合的。对象包含存储在对象内实例变量中的值。对象的主体也包含操作该对象的代码体。这些代码体称为方法。一个对象可以访问另一个对象的数据的唯一方法是调用另一个对象的方法。此操作被称为向对象发送消息。因此，对象方法的调用接口定义了该对象的外部可见部分。对象的内部部分实例变量和方法代码在外部不可见。结果是两个层次的数据抽象，这对于抽象（隐藏）对象的内部细节非常重要。面向对象的数据模型还提供对象引用，可用于标识（引用）对象。

在基于记录的模型中，数据库是由几种类型的固定格式记录构成的。每个记录都有一组固定的字段。三种最广泛接受的基于记录的数据模型是关系模型、网络模型和层次模型。后两者曾经被广泛使用过，但其重要性正在下降。关系模型的应用非常广泛。基于关系模型的数据库称为关系数据库。

关系模型使用一组表（称为关系）来表示数据和这些数据之间的关系。每个表有多个列，每个列都有一个唯一的名称。表的每一行称为一个元组，每一列表示该元组的一个属性的值。

数据库的大小变化很大，从个人数据库的几兆字节到大型企业数据库的千兆字节（千兆字节是1 000兆字节）甚至是兆字节（万亿字节是1 000兆字节）。

数据库中的信息存储在可容纳大量数据的非易失性介质上；最常用的此类介质是磁盘。磁盘可以存储比主存大得多的数据，每单位数据的成本要低得多。

为了提高关键任务系统的可靠性，可以将磁盘组织成通常称为独立磁盘冗余阵列（RAID）的结构。在RAID系统中，数据在多个磁盘上以一定数量的冗余（如复制）进行组织。即使RAID系统中的一个磁盘损坏并丢失数据，也可以从RAID系统中的其他磁盘重建丢失的数据。参见计算机存储技术。

逻辑上，关系数据库中的数据被组织为一组关系，每个关系由一组记录组成。这是给数据库用户的视图。磁盘上的底层实现（对用户隐藏）由一组文件组成。每个文件由一组固定大小的磁盘存储器组成，称为块。关系的记录存储在块中。每个关系都与一个或多个文件关联。一般来说，一个文件只包含一个关系中的记录，但出于性能考虑，如果某个文件包含多个关系的记录，则也会使用该组织。

在关系数据库中检索所需记录的一种方法是对相应的关系执行扫描；扫描从关系中提取所有记录，一次一个。

使用对关系的扫描从大型关系访问所需的记录可能非常昂贵。索引是允许更有效地访问记录的数据结构。索引建立在关系的一个或多个属性上；这些属性构成搜索键。给定每个搜索键属性的值，索引结构可用于快速检索具有指定搜索关键字值的记录。索引还可以支持其他操作，例如获取搜索键值在指定值范围内的所有记录。

数据库模式由数据定义语言表示的一组定义指定。执行数据定义语言语句的结果是存储在称为数据字典的特殊文件中的一组信息。数据字典包含元数据，即关于数据的数据。在数据库系统中读取或修改实际数据之前，会查阅此文件。数据定义语言还用于指定存储结构和访问方法。

数据操作是对存储在数据库中的信息进行检索、插入、删除和修改。数据操作语言允许用户访问或操作由适当的数据模型组织的数据。基本上有两种类型的数据操作语言：过程数据操作语言要求用户指定需要哪些数据以及如何获取这些数据；非过程数据操作语言要求用户指定需要哪些数据，而不指定如何获取这些数据。

查询是请求检索信息的语句。数据操作语言中涉及信息检索的部分称为查询语言。虽然在技术上不正确，但通常的做法是将术语查询语言和数据操作语言同义。

数据库语言支持数据定义和数据操作功能。虽然已经提出并实现了许多数据库语言，但SQL已经成为大多数关系数据库系统支持的标准语言。基于面向对象模型的数据库还支持类似于SQL的声明式查询语言。

SQL提供了一种完整的数据定义语言，包括创建与指定属性类型的关系的能力，以及对数据定义完整性约束的能力。

示例查询（QBE）是一种用于指定查询的图形语言。它广泛应用于个人数据库系统中，因为对于非专业用户来说，它比SQL简单得多。

表单界面显示一个类似于表单的屏幕视图，其中包含由用户填写的字段。有些字段可以由表单系统自动填充。报表编写器允许定义报表格式，以及从数据库中获取数据的查询；查询结果在报表中以格式显示。这些工具实际上为构建数据库接口提供了一种新的语言，通常被称为第四代语言（4GLs）。参见人机交互。

通常，数据库上的几个操作形成一个逻辑工作单元，称为事务。交易的一个例子是资金从一个账户转到另一个账户。数据库中的事务反映了商业世界中相应的事务。

传统上，数据库系统被设计为支持商业数据，主要由结构化的字母数字数据组成。近年来，数据库系统增加了对一些非传统数据类型的支持，例如文本文档、图像、地图和其他空间数据。目标是使数据库成为通用服务器，可以存储所有类型的数据。供应商并没有在核心数据库中添加对所有此类数据类型的支持，而是提供与数据库集成以提供此类功能的附加包。

Fast Reading Two Database Security

Database security concerns the use of a broad range of information security controls to protect databases (potentially including the data, the database applications or stored functions, the database systems, the database servers and the associated network links) against compromises of their confidentiality, integrity and availability. It involves various types or categories of controls, such as technical, procedural/administrative and physical. Database security is a specialist topic within the broader realms of computer security, information security and risk management.

Fast Reading Two

Security risks to database systems include:

Unauthorized or unintended activity or misuse by authorized database users, database administrators, or network/systems managers, or by unauthorized users or hackers (e.g. inappropriate access to sensitive data, metadata or functions within databases, or inappropriate changes to the database programs, structures or security configurations).

Malware infections causing incidents such as unauthorized access, leakage or disclosure of personal or proprietary data, deletion of or damage to the data or programs, interruption or denial of authorized access to the database, attacks on other systems and the unanticipated failure of database services.

Overloads, performance constraints and capacity issues resulting in the inability of authorized users to use databases as intended.

Physical damage to database servers caused by computer room fires or floods, overheating, lightning, accidental liquid spills, static discharge, electronic breakdowns/equipment failures and obsolescence.

Design flaws and programming bugs in databases and the associated programs and systems, creating various security vulnerabilities (e.g. unauthorized privilege escalation), data loss/corruption, performance degradation etc.

Data corruption and/or loss caused by the entry of invalid data or commands, mistakes in database or system administration processes, sabotage/criminal damage etc.

Ross J. Anderson has often said that by their nature large databases will never be free of abuse by breaches of security; if a large system is designed for ease of access it becomes insecure; if made watertight it becomes impossible to use. This is sometimes known as Anderson's Rule.

Many layers and types of information security control are appropriate to databases, including:

Access control

Auditing

Authentication

Encryption

Integrity controls

Backups

Application security

Database Security applying Statistical Method

Databases have been largely secured against hackers through network security measures such as firewalls, and network-based intrusion detection systems. While network security controls remain valuable in this regard, securing the database systems themselves, and the programs/functions and data within them, has arguably become more critical as networks are increasingly opened to wider access, in particular access from the Internet. Furthermore, system, program, function and data access controls, along with the associated user identification, authentication and rights management functions, have always been important to limit and in some cases log the activities of authorized users and administrators. In other words, these are complementary approaches to database security, working from both the outside-in and the inside-out as it were.

Many organizations develop their own "baseline" security standards and designs detailing basic security control measures for their database systems. These may reflect general information security requirements or obligations imposed by corporate information security policies and applicable laws and regulations (e.g. concerning privacy, financial management and reporting systems), along with generally accepted good database security practices (such as appropriate hardening of the underlying systems) and perhaps security recommendations from the relevant database system and software vendors. The security designs for specific database systems typically specify further security administration and management functions (such as administration and reporting of user access rights, log management and analysis, database replication/synchronization and backups) along with various business-driven information security controls within the database programs and functions (e.g. data entry validation and audit trails). Furthermore, various security-related activities (manual controls) are normally incorporated into the procedures, guidelines etc. relating to the design, development, configuration, use, management and maintenance of databases.

.End.

参考译文 快速阅读2 数据库安全

数据库安全涉及使用广泛的信息安全控制措施来保护数据库（可能包括数据、数据库应用程序或存储函数、数据库系统、数据库服务器和相关的网络链接）不受其机密性、完整性和可用性的损害。它涉及各种类型或类别的控制，如技术、程序/管理和物理控制。数据库安全是计算机安全、信息安全和风险管理等领域的一个专业课题。

数据库系统的安全风险包括：

授权数据库用户、数据库管理员或网络/系统管理员，或未经授权的用户或黑客未经授权或无意的行为或滥用（例如，对数据库内敏感数据、元数据或功能的不当访问，或对数据库程序、结构或安全配置的不当更改）。

恶意软件感染，导致未经授权的访问、泄露或泄露个人或专有数据、删除或损坏数据或程序、中断或拒绝授权访问数据库、攻击其他系统和数据库服务意外故障等事件；

过载、性能限制和容量问题导致授权用户无法按预期使用数据库。

由于计算机房火灾或洪水、过热、闪电、意外液体溢出、静电放电、电子故障/设备故障和过时而对数据库服务器造成的物理损坏。

数据库及相关程序和系统中的设计缺陷和编程缺陷，造成各种安全漏洞（如未经授权的权限提升）、数据丢失/损坏、性能下降等。

由于输入无效数据或命令、数据库或系统管理过程中的错误、破坏/刑事破坏等而导致的数据损坏和/或丢失。

Ross J. Anderson经常说，大型数据库的本质永远不会因为安全漏洞而被滥用；如果一个大型系统的设计是为了便于访问，那么它就变得不安全；如果做到了无懈可击，它就不可能使用了。这有时被称为安德森法则。

许多层次和类型的信息安全控制适用于数据库，包括：访问控制；审计；身份验证；加密；完整性控制；备份；应用程序安全性；应用统计方法实现数据库安全。

通过防火墙和基于网络的入侵检测系统等网络安全措施，数据库在很大程度上受到黑客的保护。虽然网络安全控制在这方面仍然很有价值，但随着网络对更广泛的访问，特别是从互联网的访问，保护数据库系统本身以及其中的程序/功能和数据的安全无疑变得更加重要。此外，系统、程序、功能和数据访问控制，以及相关的用户识别、身份验证和权限管理功能，对于限制和在某些情况下记录授权用户和管理员的活动一直很重要。换句话说，这些是数据库安全性的补充方法，从外到内和从内到外都是如此。

许多组织制定自己的“基准”安全标准，并设计详细说明数据库系统的基本安全控制措施。这些可能反映了公司信息安全政策和适用法律法规（如隐私、财务管理和报告系统）规定的一般信息安全要求或义务，以及公认的良好的数据库安全实践（例如对底层系统进行适当的强化）以及相关数据库系统和软件供应商的安全建议。特定数据库系统的安全设计通常规定进一步的安全管理和管理功能（例如管理和报告用户访问权限、日志管理和分析，数据库复制/同步和备份）以及数据库程序和功能中的各种业务驱动的信息安全控制（例如数据输入验证和审计跟踪）。此外，各种与安全有关的活动（手动控制）通常被纳入与数据库的设计、开发、配置、使用、管理和维护有关的程序、指南等。

Exercises

Ex 1 What is Database?? Try to give a brief summary according to the passage.

Ex 2 Fill in the table below by matching the corresponding Chinese or English equivalents.

database	
	关系型数据库
model	
	数据库管理系统
update	
	对象型数据库

Ex 3 Choose the best answer to the following statements according to the text we've learnt .

1. A ________ is an organized collection of data.
   A. interfaces　　B. programs
   C. database　　D. account numbers
2. Formally, a "database" refers to a set of ________ and the way it is organized.
   A. platform　　B. related data
   C. word　　D. information
3. A database ________ is a type of data model that determines the logical structure of a database and fundamentally determines in which manner data can be stored, organized, and manipulated.
   A. model　　B. Query language
   C. SQL　　D. API
4. The most popular example of a database model is ________.
   A. platform　　B. the relational model
   C. Media Player　　D. program
5. A ________ is a computer software application that interacts with the user, other applications, and the database itself to capture and analyze data.
   A. DBMS　　B. Unix
   C. Linux　　D. 3DMAX

Part B Practical Learning

Task One

Task Two

Training Target

In this part, there are two tasks in English environment. You should complete these tasks in groups under the joint guidance of professional teachers and laboratory teachers, so as to train and improve your ability to complete professional tasks in English environment.

In this part, students must finish two special tasks in English environment. Under the guidance of the specialized English teacher, the students can create data information and database for their goods. First the teacher must divide the students into several groups.

Task One Screening the Information of Goods and learning Website Design

In this task, students have two small tasks, one is to screen useful commodity information, and the other is to learn to design the establishment of online shopping mall database.

First:filter useful commodity information

Filter useful commodity information, such as pictures, fabrics, prices, places of origin, etc.

Second: Students learn the establishment of online shopping mall database.

The following is the method and steps to establish the online shopping mall database for students' reference.

Database design is usually divided into six stages

The first step:

Requirement analysis: analyzing user requirements, including data, function and performance requirements.

The second step:

Conceptual structure design: mainly using E-R model for design, including drawing E-R diagram.

The third step:

Logical structure design: by transforming E-R diagram into table, the transformation from E-R model to relational model is realized.

The fourth step:

Physical design of database: mainly choose the appropriate storage structure and access path for the designed database.

The fifth step:

Database implementation: including programming, testing and trial operation.

The sixth step:

Database operation and maintenance: system operation and daily maintenance of database.

Task Two Create Database for the Goods

In this task, students try to create a database for their online store. The following is an example of online shopping mall database design for students' reference, so as to help students think about how to establish their own shopping mall database system.

Online shopping mall database generally includes five modules:

User module:

Including attributes: user name, password, telephone, email, ID number, address, name, nick name.

Optional unique identity attributes: user name, ID card, telephone.

Commodity module:

Including attributes: commodity code, commodity name, commodity description, commodity category, supplier name, weight, validity period, price.

Order module:

Including attributes: Order number, user name, user phone, receiving address, commodity number, commodity name, quantity, price, order status, payment status, order type.

Shopping cart module:

Including attributes: user name, commodity number, commodity name, commodity price, commodity description, commodity classification, adding time, commodity quantity commodity.

Supplier module:

Including attributes: supplier number, supplier name, contact person, telephone number, business license number, address, legal person.

Students can add and delete according to their actual situation.

Part C Occupation English

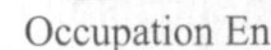
Occupation English

In this part, there is an English dialogue in real life and work environment. You will play the roles of A and B and read the dialogue aloud to practice your ability to use English.

How to Install a Database?

如何安装数据库？

Role settings: Microsoft technical service personnel (A), a company database administrator (B)

角色设置：微软技术服务人员（A），公司数据库管理员（B）

A: Hello, Microsoft customer service center. What can I do for you? 你好，微软客户服务中心。我能为你做什么？

B: Hello, Our company wants to install a database management system software. 你好，我们公司想安装一款数据库管理系统软件。

A: OK, which database software do you want to install? Is it simple Access database management software or professional relational database SQL? 好，那决定要安装哪款数据库软件？是简单的Access 数据库管理软件，还是专业点的关系型数据库SQL？

B: We want to install the SQL database management system. 我们想要安装下SQL这个数据库管理系统。

A: First, you should download the SQL server installation file from the official website. After the download is complete, locate the installation file and double-click the. Exe installation file. In the setup dialog box, select the installation option on the left and click the “new SQL Server stand-alone installation or add features to an existing installation” option on the right. n the “setup program” dialog box, the interface appears. Click OK to enter the next step, and then follow the prompts step by step. Finally, don’t forget to set the password. 首先你应该在官网上下载SQL Server安装文件。下载完成后，找到这个安装文件，并且双击.exe安装文件，进入SQL安装程序。在安装程序的对话框中，选择左边的安装选项，单击右边的“全新SQL Server独立安装或向现有安装添加功能”选项。在打开的“安装程序”对话框中，单击“确定”按钮，进入下一步操作，然后按照提示一步一步来就可以了，最后别忘了设置密码。

B: OK, I’ll try to install it. If I don’t understand, I’ll ask you again. 好的，我试着安装看看，有什么不明白的再请教。

A: Good, welcome to inquire at any time. I’m glad to serve you. 好的，欢迎随时来电咨询，很高兴为您服务。

Word Building

前缀/后缀由一个或几个字母组成，放在词根或单词之前/之后，组成一个新词。

(1) -ize (-ise)：使成为……、变成……状态、……化

dramatic 戏剧的 —————— dramatize 改编成剧本

modern 现代的—————— modernize 现代化

(2) -ent：具有……性质的、关于……的

Depend 依赖 —————— dependent 依赖的

appear 出现 —————— apparent 明显

(3) re-：回、向后

turn 转 —————— return 回来，返回

call 打电话 —————— recall 回忆，召回

(4) ex-（前缀）：前任的，以前的

president 总统 —————— ex-president 前任总统

soldier 军人 —————— ex-soldier 退伍军人

(5) de-：向下

plane 飞机 —————— deplane 下飞机

throne 王位 —————— dethrone 使离王位

Ex Translate the following words and try your best to guess the meaning of each word on the right according to the clues given on the left.

value	价值（名词）	devalue	________
press	按压 （动词）	depress	________
port	港口（名词）	export	________
tract	拖（动词 ）	extract	________
husband	丈夫（名词）	ex-husband	________
frequency	频率（名词）	frequent	________
organ	组织（名词）	organize	________
obey	服从（动词）	obedient	________

Exercises

Ex 1 What is Database? Try to give a brief summary of this passage in no more than five sentences.

Ex 2 How many DataBase model are there? Try to give several examples.

Ex 3 What is DBMS?

Ex 4 Fill in the table below by matching the corresponding Chinese or English equivalents.

DBMS	
	集合
Data Base	
	数据
model	
	结构
data processing	
	互通
SQL	
	导航数据库

Ex 5 Choose the best answer for each of the following statements according to the text we've learnt.

1. ________ is the collection of schemas, tables, queries, reports, views, and other objects.

   A. Tools B. Database

   C. Symbols D. Instructions

2. A database is an organized collection of ________.

   A. data B. information

   C. software D. instruction

3. Sometimes a DBMS is loosely referred to as a "________".

   A. operation system B. database

   C. office D. data

4. A database model is a type of ________ that determines the logical structure of a database and fundamentally determines in which manner data can be stored, organized, and manipulated.

   A. data model B. Query language

   C. SQL D. API

5. ________ is an object model language standard.

   A. OQL B.SOL

   C.FDDI D.XML

Project Seven

Online Shopping Mall's Security

Part A Theoretical Learning

Part B Practical Learning

Part C Occupation English

Part A Theoretical Learning

Training Target

In this part, our target is to improve the speed of reading professional articles and the comprehension ability of the reader. We have marked specialized key words and some flexible sentences. Try to grasp the main idea of each paragraph.

Skill One Computer Virus

What is a computer virus? **A computer virus is a special kind of computer program that reproduces its own code by attaching itself to other executable files, and spreads usually across disks and networks surreptitiously.**

Viruses have many different forms, but they all potentially have two phases to their execution: the **infection phase** and the **attack phase**.

Infection Phase

The virus has the potential to infect other programs when it executes. We often don't clearly understand when the virus will infect the other programs. Some viruses infect other programs each time they are executed; other viruses infect only upon a certain **trigger**. This trigger could be anything: a day or a time, an external event on your PC, a counter within the virus, etc. Virus writers want their programs to spread as far as possible before anyone notices them.

It is a serious mistake to execute a program a few times — find nothing infected and presume there are no viruses in the program. Maybe the virus simply hasn't yet triggered its infection phase!

Many viruses go resident in the memory of your PC in the same or similar way as **Terminate and Stay Resident(TSR) Programs**. This means the virus can wait for some external event before it infects additional programs. The virus may silently **lurk** in memory waiting for you to access a diskette, copy a file, or execute a program before it infects anything. This makes viruses more difficult to analyze since it's hard to guess what trigger condition they use for their infection.

virus ['vaɪrəs] *n.* 病毒

surreptitiously [ˌsʌrəp'tɪʃəsli] *adv.* 秘密地

infection phase 感染期

attack phase 攻击期

infect [ɪn'fekt] *vt.* 传染，感染

trigger ['trɪgə(r)] *n.* 触发事件

Terminate and Stay Resident Programs 终止驻留程序

lurk [lɜːk] *v.* 潜伏

Resident viruses frequently take over portions of the system software on the PC to hide their existence. This technique is called stealth. **Polymorphic** techniques also help viruses infect yet avoid detection.

resident virus 驻留的病毒

polymorphic [ˌpɒlɪ'mɔːfɪk] *adj.* 多态的

It's noted note that **worms** often take the opposite approach to spread as fast as possible. While this makes their detection virtually certain, they also have the effect of bringing down networks and denying access to the network. This is one of the goals of many worms.

worm [wɜːm] *n.* 蠕虫

During the infection phase, in order to infect a computer, a virus has to have the chance to execute its codes. Viruses usually ensure that this happens by behaving like a **parasite**, that is, by modifying other items so that the virus codes are executed when the legitimate items are run or opened.

parasite ['pærəsaɪt] *n.* 寄生虫

Good vehicles for viruses include the parts of a disk which contain codes executed. As long as the virus is active on the computer, it can copy itself to files or disks that are accessed.

Skill One

Viruses can be transmitted by:

△ booting a PC from an infected medium

△ executing an infected program

△ opening an infected file

Common routes for virus **infiltration** include:

△ floppy disks or other media that users can exchange

△ **E-mail attachment**

△ **pirated software**

△ shareware

infiltration [ˌɪnfɪl'treɪʃn] *n.* 渗入

E-mail attachment 邮件附件

pirated software 盗版软件

Attack Phase

Many viruses do unpleasant things such as deleting files or changing random data on your disk or merely slowing your PC down; some viruses do less harmful things such as playing music or creating messages or animation on your screen. Just as the infection phase can be triggered by some events, the attack phase also has its own trigger.

delete file 删除文件

Does this mean a virus without an attack phase is **benign**? No. Most viruses have **bugs** in them and these bugs often cause unintended negative side effects. In addition, even if the virus is perfect, it still steals system resources.

benign [bɪ'naɪn] *adj.* 良性的

bug [bʌg] *n.* 漏洞

Viruses often delay revealing their presence by launching their attack only after they have had **ample** opportunity to spread. This means the attack could be delayed for days, weeks, months, or even years after the first **initial** infection.

ample ['æmpl] *adj.* 充足的

initial [ɪ'nɪʃl] *adj.* 开始的，最初的

The attack phase is optional. Many viruses simply reproduce and have no trigger for an attack phase. Does this mean these are "good" viruses? No! Anything that writes itself to your disk without your permission is stealing storage and CPU **cycles**. What is worse is that viruses of "just infect" without an attack phase, often damage the programs or disks they infect. This is not an intentional act of virus, but simply a result of the fact that many viruses contain extremely poor defective codes.

cycle ['saɪkl] *n.* 周期

.End.

Key Words

virus *n.* 病毒

surreptitiously *adv.* 秘密地

infection phase 感染期

attack phase 攻击期

infect *vt.* 传染，感染

trigger *n.* 触发事件

Terminate and Stay Resident Programs 终止驻留程序

lurk *v.* 潜伏

resident virus 驻留的病毒

polymorphic *adj.* 多态的

worm *n.* 蠕虫

paresite *n.* 寄生虫

infiltration *n.* 渗入

E-mail attachment 邮件附件

pirated software 盗版软件

delete file 删除文件

benign *adj.* 良性的

bug *n.* 漏洞

ample *adj.* 充足的

initial *adj.* 开始的，最初的

cycle *n.* 周期

参考译文 技能1 计算机病毒

什么是计算机病毒？计算机病毒是以把自身附加到可执行文件的方式来复制其自身代码的特殊的计算机程序，并常常利用磁盘和网络秘密地进行传播。

病毒具有多种不同的形式，但是它们在执行中可能都要有两个时期：感染期和攻击期。

感染期

病毒在执行时可能会感染其他程序。我们通常无法清楚了解病毒究竟要在何时感染其他程序。一些病毒在每一次执行时都会感染其他程序，另一些病毒只有当某一个触发事件发生时才能进行传染。这种触发事件可能是任何事件：一个日期或某一个时间，一个计算机的外部事件，一个病毒内部的计数器等。病毒的编写者希望他们的程序能在其他人发现之前尽可能地广

泛传播。

对一个程序执行了几次，发现没有受到感染，就认为程序里没有病毒，这种想法是一种严重的错误。也许只是病毒还没有进入它的感染期！

许多病毒驻存在电脑的内存中，其方式与终止驻留程序相同或相似。这就意味着病毒可以等待某个外部事件的发生，然后再感染其他程序。病毒可以默默地潜伏在内存中，等你访问一个磁盘、复制一个文件或执行一个程序之后，才开始进行感染。这使病毒更加难以分析，因为很难判断病毒要借由什么样的触发条件来进行感染。

驻留的病毒经常利用部分电脑系统软件来隐藏它们的存在。这种技术被称为隐蔽。多形态技术也有助于病毒进行感染而不被发现。

需要注意的是蠕虫病毒经常采用相反的方法尽可能快地进行传播。尽管一定能够发现它们，但该病毒在被发现前已经导致了网络速度的下降，并阻止了对网络的访问。这是许多蠕虫病毒的目标之一。

在感染期，为了要传染某台计算机，病毒必须得有运行其代码的机会。为了运行代码，病毒通常会像寄生虫一样，通过修改其他程序，在合法程序运行或打开时运行病毒代码。

实现病毒传染的常见工具包括可执行代码的部分磁盘。只要病毒在计算机上是活跃的，它就能将其自身复制到计算机所访问的其他文件或磁盘上。

病毒可通过以下方式传输：

△ 通过已被感染的媒体启动计算机

△ 运行被感染的程序

△ 打开被感染的文件

病毒渗入的常见途径包括：

△ 用户间能交换的盘片或其他媒体

△ 邮件附件

△ 盗版软件

△ 共享软件

攻击期

许多病毒做一些令人十分不愉快的事情，如删除文件，或任意修改盘上的数据，或仅仅使电脑的运行速度变慢；某些病毒做一些危害性较小的事情，如播放音乐或在屏幕上显示信息或动画。就像需要某个事件来触发感染期一样，攻击期也需要自己的触发事件。

这是否意味着没有攻击阶段的病毒就是良性的呢？非也！大多数病毒都有漏洞，这些漏洞经常会在无意中引起负面效果。另外，即使病毒没有漏洞，它仍然会窃取系统资源。

病毒经常在有充分的机会来进行传播后，才进行攻击，这样就可以延迟病毒的暴露。这意味着病毒在第一次感染后，可能要延迟几天、几周、几个月甚至几年才进行攻击。

攻击期是随意的，许多病毒仅仅是繁殖，而没有攻击期的触发事件。这是否表明这些病毒就是“好”病毒呢？非也！任何未经允许就将自己写入磁盘的病毒，都在窃取存储空间和CPU周期。更糟的是，这些“仅仅感染”而没有攻击期的病毒经常破坏它们所感染的程序或磁盘。这不是病毒的故意行为，而仅仅是因为许多病毒包含了有极差缺陷的代码。

Skill Two Computer System Security Measures

There is an urgent need for computer security. Computer owners must take measures to prevent theft and **inappropriate** use of their equipment. One aspect of computer security is the protection of information against inappropriate manipulation, destruction or **disclosure**. Another security problem concerns the protection of the computer system and data on the computer. It is essential that security measures protect all operating systems.

Computer security is concerned with protecting information, hardware, and software. Security measures consist of **encryption**, **restricting access**, **data layering**, and **backing up data**.

Encryption

Encryption provides secrecy for data. Since data of encryption that cannot be read generally and also cannot be changed, encryption can be used to achieve **integrity**. Encryption is an important tool in computer security, but one should not **overrate** its importance. Users must understand that encryption does not solve all computer security problems. Furthermore, if encryption is not used properly, it can have no effect on security or can degrade the performance of the entire system. Thus, it is important to know the situations in which encryption is useful and how to use it effectively.

The most powerful tool in providing computer security is coding. It is **unintelligible** for the outside observer to transform data; the value of an **interception** and the possibility of a **modification** or a **fabrication** are almost **mollified**.

Restricting Access

Everyone in the organization needs access to the Internet. Employees should be trained on the threats that computer viruses can **impose on** information systems. They should realize how important it is not to open unidentified files, leave computer terminals on while they are away from their desks, display their passwords for anyone to see, etc.

Security experts are constantly **devising** new ways to protect computer systems from access by unauthorized persons. Sometimes security is a matter of placing guards in company computer rooms and checking the **identification** of everyone admitted. The passwords are secret words or numbers that must be keyed into a computer system to gain access.

inappropriate [ˌɪnə'prəʊpriət] *adj.* 不适当的

disclosure [dɪs'kləʊʒə(r)] *n.* 泄露

encryption [ɪn'krɪpʃn] *n.* 加密

restricting access 限制访问

data layering 数据分层

backing up data 备份数据

integrity [ɪn'tegrəti] *n.* 完整

overrate [ˌəʊvə'reɪt] *v.* 过高评价

unintelligible [ˌʌnɪn'telɪdʒəbl] *adj.* 难以理解的

interception [ˌɪntə'sepʃn] *n.* 截取

modification [ˌmɒdɪfɪ'keɪʃn] *n.* 更改，修改

fabrication [ˌfæbrɪ'keɪʃn] *n.* 伪造

mollify ['mɒlɪfaɪ] *v.* 平息

impose on 占……便宜；利用

devise [dɪ'vaɪz] *v.* 设计

identification [aɪˌdentɪfɪ'keɪʃn] *n.* 身份证明

Most major corporations today use special hardware and software called **firewalls** which act as a security **buffer** between the corporation's private network and all external networks, including the Internet to control the computer network access. All electronic communications coming into and leaving the corporation must be evaluated by the firewall. Security is maintained by **denying** access to unauthorized communications.

firewall ['faɪəwɔ:l] *n.* 防火墙
buffer ['bʌfə(r)] *n.* 缓冲，缓冲区
deny [dɪ'naɪ] *v.* 否认，拒绝

Data Layering

Organizations need to determine and classify important information, such as the company's financial information, employee data, etc. and establish different layers of security. Only certain employees should have access to this classified information.

Backing Up Data

Equipment can always be replaced. A company's data, however, may be **irreplaceable**. Most companies should take some effective measures of trying to keep software and data from being tampered with in the first place. These measures include careful screening job applicants, guarding passwords, and **auditing** data and programs from time to time. The safest procedure, however, is to make data backups frequently and to store them in remote locations.

Skill Two

irreplaceable [ˌɪrɪ'pleɪsəbl] *adj.* 不可替代的
audit ['ɔ:dɪt] *v.* 审计

From this discussion, it should be evident that how important computer security is. Computer security will continue to be an issue we need to focus on because the number of computers and users continues to grow.

End.

Key Words

inappropriate *adj.* 不适当的
encryption *n.* 加密
data layering 数据分层
integrity *n.* 完整
unintelligible *adj.* 难以理解的
modification *n.* 更改，修改
mollify *v.* 平息
devise *v.* 设计
firewall *n.* 防火墙
deny *v.* 否认，拒绝
audit *v.* 审计
disclosure *n.* 泄露
restricting access 限制访问
backing up data 备份数据
overrate *v.* 过高评价
interception *n.* 截取
fabrication *n.* 伪造
impose on 占……便宜；利用
indentification *n.* 身份证明
buffer *n.* 缓冲，缓冲区
irreplaceable *adj.* 不可替代的

参考译文 技能2 计算机系统安全措施

计算机安全已成为急需解决的问题。计算机拥有者必须采取措施防止他们的设备被窃取和非法使用。计算机安全的一个方面就是要保护信息不被越权操作、破坏或泄露。另一个安全问题涉及计算机上操作系统和数据的保护。采用安全措施保护所有的计算机操作系统是很有必要的。

计算机安全与信息、硬件和软件的保护有关。安全措施包括加密、限制访问、数据分层和备份数据。

加密

加密提供数据保密。加密的数据一般不能读出，也不能更改，因此能保证数据的完整。加密是计算机安全的重要工具，但有时也不能对它的重要性估计过高。用户一定要知道加密并不能解决计算机所有的安全问题。甚至如果加密使用不当，不但对安全没有作用，还会降低整个系统的性能。所以，重要的是要了解在什么情况下加密有用和有效。

保证计算机安全的最有效的方法是编码。对外界而言，转化数据是无规律的，这样截取的数据就是没有用处的，几乎不存在修改或伪造的可能性。

限制访问

公司内的每个人都需要访问因特网。每位职员都应在计算机病毒可利用信息系统造成威胁这一方面将要培训。他们应该明白，不要打开未经确认的文件，当他们离开办公桌时不要让电脑终端继续工作，不要让其他任何人看到自己的密码等，这些都是很重要的。

安全专家不断设计新方法，用以保护计算机系统免受未经授权的人的访问。有时，为保证安全，会派警卫看护公司计算机室，检查每个进入的人的身份证明。口令是秘密的单词或数字，必须将其键入计算机系统才能进行访问。

今天，大多数大公司都使用被称为防火墙的专门的硬件和软件来控制计算机网络的访问。这些防火墙，在公司专用网络与包括因特网在内的所有外部网络之间，起到安全缓冲区的作用。所有进出公司的电子通信都必须经过防火墙的评估。通过拒绝未经授权的通信进出来维护公司的网络安全。

数据分层

公司需要确定哪些是重要信息并将其进行分类。比如公司的财政信息、职员数据等，并建立不同的安全层次。只有特定的职员能访问这些分类信息。

备份数据

设备随时可以被替换。然而，一个公司的数据可能是无法替代的。因此，大多数公司应该在第一时间采取一些有效的措施，防止软件和数据被篡改。方法包括仔细审查求职者、严守口令，以及时常检查数据和程序。然而，最保险的办法是经常制作数据备份，并将其存放在远距离地点。

通过以上讨论，我们应当清楚计算机安全的重要性。随着计算机和计算机用户人数的不断增加，计算机安全将会继续成为需要人们关注的问题。

Fast Reading One A Brief Introduction to Firewalls

Firewalls are frequently used to prevent unauthorized Internet users from accessing private networks connected to the Internet. A small home network has many of the same security issues as a large corporate network does. You can use a firewall to protect your home network and family from offensive websites and potential hackers. Firewalls can be implemented in both hardware and software, or a combination of both. A firewall is simply a program or hardware device that filters the information coming through the Internet connection into your private network or computer system. All messages entering or leaving the Internet will pass through the firewall, which examines each message and blocks those that do not meet the specified security criteria. If an incoming packet of information is flagged by the filters, it is not allowed to pass through.

Typically, a firewall is placed on the entry point to a public network such as the Internet. It is a protective system that lies between your computer and the Internet. It should be considered as a traffic cop. The firewall's role is to ensure that all communication between an organization's network and the Internet conforms to the organization's security policies. Its job is similar to a physical firewall that keeps a fire from spreading from one area to the next. Primarily these systems are TCP/IP-based, and depending on the implementation, can enforce security roadblocks as well as provide administrators with answers to the following questions:

△ Who's been using network?

△ What were they doing on network?

△ When were they using network?

△ Where were they going on network?

△ Who failed to enter the network?

Fast Reading One

End.

参考译文 快速阅读1 防火墙简介

防火墙经常用于防止未经许可的因特网用户访问连接到因特网上的个人网络。小型家庭网有着许多与大公司的网络相同的安全问题。防火墙可以保护你的家庭网和家庭免遭恶意网站和潜在黑客的攻击。防火墙可以由硬件、软件或二者联合实现。一个防火墙就是一个程序或者一台硬件设备，用于过滤通过因特网连接进入你的专用网或计算机系统中的信息。进入或由内部网发出的所有信息都要经过防火墙，它检查每一个不符合指定安全标准的信息。如果一个输入的信息包被过滤器做了标记，它就不被允许通过。

典型的防火墙被置于公共网络（如因特网）入口处。它是位于你的计算机与因特网之间的保护系统。它可以被看作是交通警察。防火墙的作用是确保一个单位的网络与因特网之间的所有通信都符合单位的安全策略。它的作用类似于一堵防止火灾从一处蔓延到另一处的实实在在的防火墙。这些系统基本上遵循TCP/IP协议，并与实现方法有关，它们能实施安全路障并为管理人员提供下列问题的答案：

△ 谁在使用网络？
△ 他们在网上做什么？
△ 他们什么时间使用网络？
△ 他们去了网络的什么地方？
△ 谁要上网但没有成功？

Fast Reading Two Firewall Techniques

Fast Reading Two

In general, there are three types of firewall implementations, some of which can be used together to create a more secure environment. These implementations are packet filtering, application proxies, and circuit-level or generic-application proxies. Packet filtering is often achieved in the router itself. Application proxies, on the other hand, usually run on standalone servers. Proxy services take a different approach than packet filtering, using a modified client program that connects to a special intermediate host that actually connects to the desired service.

There are several types of firewall techniques:

Application Gateway

The first firewalls are application gateways, and are sometimes known as proxy gateways. These are made up of the bastion hosts that run special software to act as a proxy server. This software runs at the Application Layer of the OSI Reference Model and applies security mechanisms to specific applications, such as FTP and Telnet servers. This is very effective.

Packet Filter

Packet filter looks at each packet through the network and accepts or rejects it based on user-defined rules. Packet filtering is fairly effective and transparent to users, but it is difficult to configure. In addition, it is susceptible to IP spoofing.

Circuit-level Gateway

Circuit-level Gateway applies security mechanisms when a TCP or UDP connection is established. Once the connection has been made, packets can flow between the hosts without further checking.

Proxy Server

Proxy server intercepts all messages entering and leaving the network. The proxy server effectively hides the true network addresses.

Firewalls use one or more of three methods to control traffic flowing in and out of the network:

(1) Packet filtering: Packets (small chunks of data) are analyzed against a set of filters. Packets that make it through the filters are sent to the requesting system and all others are discarded.

(2) Proxy service: Information from the Internet is retrieved by the firewall and then sent to the requesting system and vice versa.

(3) Stateful inspection: It is a newer method that doesn't examine the content of each packet but instead compares certain key parts of the packet to a database of trusted information. Information traveling from inside the firewall to the outside is monitored for specific defining characteristics, and

then incoming information is compared to these characteristics. If the comparison yields a reasonable match, the information is allowed to pass through. Otherwise it is discarded.

The level of security you establish will determine how many of these threats can be stopped by your firewall. The highest level of security would be to simply block everything. Obviously that defeats the purpose of having an Internet connection. But a common rule is to block everything, and then begin to select what types of traffic you will allow. You can also restrict traffic that travels through the firewall so that only certain types of information, such as E-mails, can get through. For most of us, it is probably better to work with the defaults provided by the firewall developer unless there is a specific reason to change it.

One of the best things about a firewall from a security standpoint is that it stops anyone on the outside from logging onto a computer in your private network. While this is a big idea for businesses, most home networks will probably not be threatened in this manner.

.End.

参考译文 | 快速阅读2 防火墙技术

通常有三种类型的防火墙实现方案，有些是将几种类型一起使用，以建立一个更安全的环境。这些实现方案是：数据包过滤、应用程序代理和电路级或通用应用程序代理。数据包过滤常常是在路由器中实现的，而应用程序代理通常运行在独立的服务器上。代理服务采取不同于数据包过滤的方法，使用修改的客户端程序，该程序与专用中间主机相连，而该主机又与实际所需服务器端相连。

这里有几种不同类型的防火墙技术：

应用网关

第一种防火墙是应用网关，即平时所知的代理网关。应用网关由运行特定软件的堡垒型主机组成，作用如同代理服务器。软件运行在OSI参考模型的应用层上，并将安全机制应用于专用应用软件，比如FTP与Telnet服务器上，这是非常有效的。

数据包过滤器

数据包过滤器按照用户定义的规则，对通过网络的每一个数据包进行检查并接收或拒绝。数据包过滤是非常有效的和对用户透明的，但是设置困难。此外，它易受IP的欺骗。

电路级网关

当TCP或者UDP建立连接时，电路级网关会开启安全机制。一旦建立了连接，数据包可以在主机间流动且不需要进一步检查。

代理服务器

代理服务器截取所有进入网络和由网络发出的信息。它能有效地屏蔽真实的网络地址。

防火墙使用下列三种方法中的一种或几种来控制进出网络的通信：

（1）数据包过滤：数据包（小块数据）由一组过滤器进行分析。能通过过滤器的数据包被发送到发出请求的系统，其他的则被丢弃。

（2）代理服务：来自因特网的信息由防火墙进行检索，然后被发送到提出请求的系统，反之亦然。

（3）状态检查：这是一种更新的方法，这种方法不会检查每个数据包的内容，而是将数据包的某些关键部分与一个可信的信息数据库进行比较。从防火墙内部传输到外部的信息可根据特别规定的特性进行监控，然后系统会将输入的信息与这些特性相比较，若能生成一个合理的匹配，则允许信息通过，若不能就会丢弃信息。

你所设定的安全级别将决定你的防火墙能阻挡多少威胁。最高安全级别就是阻断一切。很显然，这就失去了进行互联网连接的意义。但通常的做法是阻断一切，然后，开始选择你将允许什么样的流量类型。你还可以限制通过防火墙的流量，以便只有几种信息可以通过，如电子邮件。对我们大多数人来说，除非有特殊的理由要改变它，否则最好在防火墙开发商提供的默认条件下工作。

从安全的角度来看，防火墙的一个优点就是它能阻止任何外来人员登录到专用网中的计算机上。尽管这一点对企业尤为重要，采用这种方式的大多数家庭网也不会受到这种威胁。

Exercises

Ex 1 **What is a computer virus? Try to describe it in a few words.**

Ex 2 **How does a computer virus infect files?**

Ex 3 **Fill in the table below by matching the corresponding Chinese or English equivalents.**

resident virus	
	病毒
worm	
	攻击期
bug	
	感染期

Ex 4 Choose the best answer for each of the following statements according to the text we've learnt.

1. A virus is a ________
    A. program
    B. computer
    C. bad man
    D. beast
2. A virus is a program that reproduces its own code by ________.(多选)
    A. adding to the end of a file
    B. inserting into the middle of a file
    C. simply placing a pointer
    D. replacing another program
3. Similar to viruses, you can also find malicious code in ________.(多选)
    A. Trojan Horses
    B. Worms
    C. Microsoft Word Documents
    D. Logic bombs
4. Viruses all have two phases to their execution, the ________ and the ________.(多选)
    A. infection phase
    B. deletion phase
    C. attack phase
    D. creation phase
5. ________ may be a certain cause that some viruses infect upon.(多选)
    A. A day
    B. A time
    C. An external event on your PC
    D. A counter within the virus
6. Many viruses go resident in the memory like a(an) ________.
    A. exe file
    B. com file
    C. TSR program
    D. data file
7. If you ________, it may wake up a virus that is resident in memory. (多选)
    A. delete a file
    B. access a diskette
    C. execute a program
    D. copy a file

8. Viruses can be transmitted by ________.

   A. booting a PC from an infected medium

   B. executing an infected program

   C. opening an infected file

   D. All of the above.

9. Common routes for virus infiltration include ________.

   A. floppy disks or other media that users can exchange

   B. E-mail attachments

   C. pirated software and shareware

   D. All of the above.

10. Although the virus simply reproduces and has no cause for an attack phase, it will still _______ without your permission.

   A. hide its own code

   B. steal storage and pilfer CPU cycles

   C. delete files

   D. play music

Part B Practical Learning

Task One

Task Two

Training Target

In this part, there are two tasks in English environment. You should complete these tasks in groups under the joint guidance of professional teachers and laboratory teachers, so as to train and improve yours ability to complete professional tasks in English environment.

Task One Discuss Potential Security Issues

In this task, students should understand the problems about the security of the online shop.

Usually there are some aspects about the security of the online shop: Firstly website is easy to be infected with viruses; secondly, website is regularly attacked by hackers. For example, customers' private information is intercepted and stolen; order information is revised; the online store's information and the customer information are fake. Other problems may include online payment security, payment repudiation and so on.

Task Two Set Up Security Measures for the Store

In this task, students can build safety measures to cope with the common online shop security issues.

These measures include:

install anti-virus software on your computer and keep it updated;

strengthen the personal identity verification;

encrypt the purchase process data;

perfect the methods of online payment;

build online store real name system and credit system.

Part C Occupation English

Occupation English

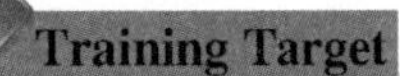

In this part, there is an English dialogue in real life and work environment. You will play the roles of A and B and read the dialogue aloud to practice your ability to use English.

Network Security 网络安全

Role Setting: Baidu Online Technical Service Center Personnel (A), A Company Network Security Officer (B)

角色设置：百度线上技术服务中心人员（A），公司网络安全办公人员（B）

A: Hello, welcome to call 360 online technical service center. What can I do for you? 您好，欢迎致电360在线技术服务中心，有什么能帮到您？

B: Hello, I am a company network security administrator. I installed an enterprise version of your anti-virus software before. I would like to consult whether there is other software or measures to ensure the security of the network. 你好，我是一名公司网络安全管理员，我之前安装了贵公司的一款企业版的防病毒软件。我想咨询下还有没有其他保障网络安全的软件或措施。

A: I see. In addition to anti-virus software, what software can guarantee network security? 明白了，你想除了防病毒软件外还有什么软件能保障网络安全是吧？

B: Yes. 是的。

A: There are many. For example, you can install an anti-spyware software. You can also install a firewall, so that the security is higher. 那有很多，比如您可以再安装一款反间谍软件，还可以安装一个防火墙，这样安全保障更高一些。

B: That's great. How can I get these two pieces of software? 那太好了，我如何获得这两款软件呢？

A: You can download them directly from our official website. 您可以在我们的官网直接下载使用。

B: Great. thank you! 太好了。谢谢！

A: You are very kind! Welcome to inquire. 您太客气了！欢迎您来电咨询。

B: Bye! 再见！

A: See you! 再见！

Word Building

前缀/后缀由一个或几个字母组成，放在词根或单词之前/之后，组成一个新词。

(1) micro-(前缀)：微，小

processor 处理器 ———— microprocessor 微处理器

(2) super-(前缀)：超

market 市场 ———— supermarket 超级市场

highway 公路 ———— superhighway 超级公路

(3) -er(后缀)：……者，……物，……的人

teach 教 ———— teacher 教师、导师

cook 烹调，煮 ———— cooker 厨具

(4) -able(后缀)：能……的，可……的，易于……的

adjust 调整 ———— adjustable 可调整的

suit 适合 ———— suitable 适当的，合适的

Ex **Try your best to guess the meaning of each word on the right according to each clues given on the left.**

computer	计算机（名词）	supercomputer	________
computer	计算机（名词）	microcomputer	________
electronic	电子的（形容词）	electron	________
		electronics	________
number	数字（名词）	numerical	________
add	加（动词）	adder	________
multiply	乘（动词）	multiplier	________
divide	分，除（动词）	divider	________
flex	折曲（动词）	flexible	________
rely	依靠（动词）	reliable	________

Exercises

Ex 1 **According to the text, if you have a system that is not currently running virus protection software, what should you do?**

Ex 2 Fill in the table below by matching the corresponding Chinese or English equivalents.

	防火墙
data layering	
	缓冲
modification	
	备份数据
restricting access	
	截取

Ex 3 Choose the best answer for each of the following statements according to the text we've learnt.

1. Security measures consist of __________.
   A. encryption
   B. restricting access
   C. data layering and backing up data
   D. All of the above.
2. The most powerful tool in providing computer security is __________.
   A. transforming data
   B. restricting access
   C. coding
   D. backing up data
3. __________ can be used to achieve integrity and secrecy for data.
   A. Restricting access
   B. Encryption
   C. Data layering
   D. Backing up data
4. It is very important that we should not __________ in protecting computer system.
   A. display our passwords for anyone to see
   B. open unidentified files
   C. leave computer terminals on while we are away from our desks
   D. All of the above.
5. All electronic communications coming into and leaving the corporation must be evaluated by the __________.
   A. passwords
   B. firewall
   C. gateway
   D. buffer

Project Eight

Let the Students Trade Online

Part A Theoretical Learning

Part B Practical Learning

Part C Occupation English

Part A Theoretical Learning

Training Target

In this part, our target is to improve the speed of reading professional articles and the comprehension ability of the reader. We have marked specialized key words and some flexible sentences. Try to grasp the main idea of each paragraph.

Skill One E-Commerce

[In the 21st century], the rapid development of information technology and the rapid increase in information exchange have brought new drives and **innovative** ideas to the whole society. The wide adoption of information technology by the community has led to great changes. These changes are not simply in the context of data processing or computing. They are changes which affect how we communicate with each other, how we organize our daily activities, how we educate the younger generation, and how we run business. The development and wide adoption of information technology, computer network and the Internet have **transformed** the mode of operation of many businesses, and at the same time have brought along **unprecedented** business opportunities. Businesses are now able to conduct transactions across geographical **boundaries**, across time zones and at a high efficiency. **E-Commerce** has become the market trend of the century.

innovative ['ɪnəveɪtɪv] *adj.* 革新的；创新的

transform [træns'fɔ:m] *vt.* 改变；改观

unprecedented [ʌn'presɪdentɪd] *adj.* 空前的，无前例的

boundary ['baʊndri] *n.* 分界线

E-commerce 电子商务

What's E-Commerce?

E-Commerce is doing business through electronic media. It means using simple, fast and low-cost electronic communications to transact, without face-to-face meeting between the two parties of the transaction. Now, it is mainly done through the Internet and Electronic Data Interchange (EDI). E-Commerce was first developed in the 1960s. With the wide use of computers, the maturity and the wide adoption of the Internet, the **permeation** of credit cards, the establishment of secure transaction agreement and the support and promotion by governments, the development of E-Commerce is

Skill One

permeation [ˌpɜ:mi'eɪʃn] *n.* 渗入，透过

becoming **prosperous**, with people starting to use electronic means as the media of doing business.

prosperous ['prɒspərəs] *adj.* 繁荣的，兴旺的

Types of E-Commerce

◇ Electronic network within the company: through the Internet, people can exchange and handle business information internally.

◇ Business-to-Business (**BTB** or B2B): [amongst all the types of E-Commerce], this way of doing business electronically through the Internet or Electronic Data Interchange is the one that deserves the most attention. As estimated by Forrester Research, BTB E-Commerce will grow at a rate three times that of Business-to-Consumer E-Commerce and thus has the greatest potential for growth.

BTB 企业对企业

◇ Business-to-Consumer(**BTC** or B2C): businesses provide consumers with online shopping through the Internet, allowing consumers to shop and pay their bills online. This type of offering saves time for both **retailers** and consumers.

BTC 企业对消费者

retailer ['ri:teɪlə(r)] *n.* 零售商，零售店

◇ Consumer-to-Consumer: consumers can post their own products online through some agent websites for other consumers to bid.

◇ Government-to-Business: this mode of trading often describes the way in which government purchases goods and services through electronic media such as the Internet, for example the Electronic Tendering System (ETS). This system is an infrastructure to provide online services such as **registration** of suppliers, tender **notification**, downloading facility for tender documents, enquiries handling, submission of tender proposals and announcement of tender results.

registration [ˌredʒɪ'streɪʃn] *n.* 注册

notification [ˌnəʊtɪfɪ'keɪʃn] *n.* 通知

.End.

Key Words

innovative *adj.* 革新的；创新的	transform *vt.* 改变；改观
unprecedented *adj.* 空前的，无前例的	boundary *n.* 分界线
E-commerce 电子商务	permeation *n.* 渗入，透过
prosperous *adj.* 繁荣的，兴旺的	BTB 企业对企业
BTC 企业对消费者	retailer *n.* 零售商，零售店
registration *n.* 注册	notification *n.* 通知

参考译文 技能1 电子商务

21世纪，信息技术的快速发展和信息交换的快速增长已经为整个社会带来了新的动力和创新思想。社会对信息技术的广泛应用已经为其带来巨大的变化。这些变化不单单是在数据处理或计算上，还包括我们如何进行交流，如何安排我们的日常活动，如何教育下一代，如何做生意。信息技术的快速发展和广泛应用以及计算机网络和因特网已经改变了许多商业的运作方式，同时也带来了前所未有的商机。现在的企业能够跨越地域的界限，穿越时空的差别，以很高的效率进行交易。电子商务已经成为21世纪市场的发展趋势。

什么是电子商务？

电子商务就是通过电子媒介来做生意。它意味着利用简单、快捷而且低廉的电子通信手段进行交易，而交易的双方不必面对面地会晤。现在，电子商务主要通过因特网和EDI进行交易。电子商务最早起源于20世纪60年代。随着计算机的广泛应用、因特网的成熟和广泛采用、信用卡的介入、安全交易协议的建立和政府的支持与推动，电子商务的发展前景越来越广阔，人们开始利用电子商务做生意。

电子商务的类型

◇ 企业内的电子网络：人们可以通过企业内联网交换和处理内部商业信息。

◇ 企业对企业的电子商务（BTB或B2B）：在电子商务的许多种类中，这种通过互联网和EDI进行交易的方式是最受关注的。据弗雷斯特研究公司的预测，BTB电子商务将以3倍于BTC电子商务的速度增长，因此有着最大的发展潜力。

◇ 企业对消费者的电子商务（BTC或B2C）：企业通过互联网为消费者提供在线购物，使消费者可以在线购买商品及付账。这种方式节省了零售商和消费者的时间。

◇ 消费者与消费者之间的电子商务：消费者可以通过代理商的网站在线发布自己的产品以供其他消费者竞买。

◇ 政府与企业之间的电子商务：这种交易方式是指政府通过因特网等电子媒介购买商品和服务，例如电子投标系统（ETS）。这个系统提供了在线服务的基础构架，如供应商的注册、投标通知、投标文件的下载、投标咨询、投标计划的递交和投标结果的发布。

Skill Two Online shopping

Online shopping or online retailing (Pic 8.1) is a form of electronic commerce which allows consumers to directly buy goods or service from a seller over the Internet using a web browser. Alternative names are: e-web-store, e-shop, e-store, Internet shop, web-shop, web-store, online store, online storefront and virtual store. Mobile commerce (or m-commerce) describes purchasing from an online retailer's mobile optimized online site or app.

online shopping 网上购物

Skill Two

Pic 8.1 Online Shopping

The major advantage of online shopping is the convenience it offers. By sitting back at home you can now shop anything from candles to vehicles by several clicks of mouse button. One of the important disadvantages of online shopping is lack of personal interaction. Another disadvantage of online shopping is tangibility factor. Seeing the picture of a product is far inferior to that of seeing it in real world. When you go for real world shopping, you can actually touch, feel or sense it with different means, but for online shopping you can only view the electronic catalogues. Even though this problem has been rectified to certain extent by use of 3D product catalogues, some online malls still use the old-fashioned images in product catalogues.

Paying by **credit card** is the widely accepted method of payment for online shopping. However, the other methods, like using e-checks, **PayPal** and **bank transfer** are also common. The method of payment is decided upon the mutual trust and familiarity between online merchant and the customer.

credit card 信用卡
PayPal 贝宝（全球最大的在线支付平台）
bank transfer 银行汇款，银行转账

Retail success is no longer all about physical stores. This is evident because of the increase in retailers now offering online store interfaces for consumers. With the growth of online shopping comes a wealth of new market footprint coverage opportunities for stores that can appropriately cater to offshore market demands and service requirements.

Online stores must describe products for sale with texts, photos, and multimedia files, whereas in a physical retail store, the actual product and the manufacturer's packaging will be available for direct inspection (which might involve a test drive, fitting, or other experimentation).

Some online stores provide or link to supplemental product

information, such as instructions, safety procedures, demonstrations, or manufacturer specifications. Some provide background information, advice, or how-to guides designed to help consumers decide which product to buy.

Online shopping is a different experience and you can make the shopping creative over the Internet as you get used to it. There can be a lot of apprehensions about online shopping when you get into it for the first time. As you experience more and more of it those apprehensions get disappeared slowly. Remember that if you stick to the basics, online shopping will become more enjoyable and easier than real-world shopping.

.End.

Key Words

online shopping 网上购物	credit card 信用卡
PayPal 贝宝	bank transfer 银行汇款，银行转账

参考译文　技能2 网上购物

网上购物或网上零售（图8.1）是电子商务的一种形式，它可以让消费者通过网络浏览器直接从卖家那里购买商品或服务。可用的名字有：电子网络商店、电子购物、电子商店、因特网购物、网络购物、网络商店、在线商店、线上店面和虚拟商店。移动电子商务一词描述了这种从在线零售商的移动优化网站或应用程序购买物品的形式。

网上购物最大的优点是方便。你坐在家里通过点击几下鼠标就可以购买从蜡烛到车在内的所有商品。在线购物的一个最大的缺点是缺乏人际交流，另一个缺点是无法真实触碰到商品。单看产品的图片远不及在实体店看实实在在的产品。当你去实体店的时候，你能通过触摸、感觉等多种不同的方法了解商品，但是在线购物时你只能浏览电子目录。虽然这个问题已经通过使用3D电子产品目录在一定程度上得到改进，但是还有很多商家依然在产品目录中使用过时的图片。

信用卡支付是网上购物应用最广泛的一种支付方式。但其他支付办法，像电子支票、贝宝和银行转账也是常见的方法。支付方法的选择是建立在网上商家和客户之间的相互信任和熟悉度基础之上的。

零售业的成功不再只看实体店，因为现在已经有越来越多的零售商为用户提供在线商店。网上购物的增长，对于商家来说，是丰富其商品覆盖率的机会，可以适当地迎合海外市场的需求和服务要求。

网上商店必须使用文字、图片和相应的多媒体文件去描述产品，而在实体零售店，真实的商品和生产商的包装更方便直接检查（包括实际检查，试用，查看是否合适及其他检验方法）。

一些网上商店会提供或者链接产品补充信息，如使用说明书、安全规则、产品展示或制造商规格。一些网店还会提供相关背景信息链接、建议或指南来帮助消费者决定购买哪种商品。

网上购物是一种不同的体验，习惯之后你能通过网络使购物变得很灵活。第一次使用网络购物时你可能会有一些担心，但当你越来越多地使用它时，这种担心就会慢慢消失。请记住，如果你遵守基本规划，网上购物要比实体店购物有趣、简单得多。

Fast Reading One An Introduction to Programming Languages

Fast Reading One

Computers cannot function without programs, which give them instructions. People specialized in writing programs are known as computer programmers. They construct programs by using programming languages.

Programming language, in computer science, is the artificial language used to write a sequence of instructions that can be run by a computer. These are not natural languages, such as English, Chinese, but are specified sets of words, phrases and symbols called codes, which can be combined in certain very restricted ways to instruct the computer.

However, natural languages are not suited for programming computers because they are ambiguous, meaning that their vocabulary and grammatical structure may be interpreted in multiple ways. The languages used to program computers must have simple logical structures, and the rules for their grammar, spelling and punctuation must be precise.

Programming languages date back almost to the invention of the digital computer in the 1940s. Computer languages have undergone dramatic evolution since the first electronic computers were built to assist in telemetry calculations during World War Ⅱ.

Early on, programmers worked with the most primitive computer instructions: machine language. These instructions were represented by long strings of ones and zeros. The first assembly languages emerged in the late 1950s with the introduction of commercial computers. It maps machine instructions to human-readable mnemonics, such as ADD and MOV.

The first procedural languages were developed in the late 1950s to early 1960s, FORTRAN created by John Backus and the COBOL created by Grace Hopper. The first functional language was LISP, written by John McCarthy in the late 1950s. Although heavily updated, all three languages are still widely used today. In the late 1960s, the first object-oriented languages, such as SIMULA, emerged. During the 1970s, procedural languages continued to develop with ALGOL, BASIC, PASCAL and C.

Programming languages use specific types of statements, or instructions, to provide functional structure to the program. A statement in a program is a basic sentence that expresses a simple idea—its purpose is to give the computer a basic instruction. Statements define the types of data allowed, how data is to be manipulated, and the ways that procedures and functions work.

Variables can be assigned different values within the program. The properties variables can have are called types. In many programming languages, a key data type is a pointer. Pointers themselves do not have values; instead, they have information that the computer can use to locate some other variable—that is, they point to another variable.

An expression is a piece of a statement that describes a series of computations to be performed on some of the program's variables, such as X + Y/Z, in which the variables are X, Y, and Z. The computations are addition and division. An assignment statement assigns a variable a value derived from some expression, while conditional statements specify expressions to be tested and then used to select which other statements should be executed next.

Procedure and function statements define certain blocks of code as procedures or functions that can then be returned later in the program. These statements also define the kinds of variables and parameters the programmer can choose and the type of value that the code will return when an expression accesses the procedure or function.

The most commonly used programming languages are highly portable and can be used to effectively solve diverse types of computing problems. Programming languages can be classified as either low-level programming languages or high-level programming languages.

Low-level programming languages, for example, assembly languages, are the most basic type of programming languages. Assembly languages are very close to machine languages, but they must still be translated into machine languages. How do you understand machine languages? Machine languages are a sequence of 1s and 0s, called bits, and can be understood directly by a computer.

In the same way, high-level programming languages are close to human natural languages, and also must first be translated into machine languages by a compiler before they can be understood and processed by a computer. Examples of high-level programming languages are FORTRAN, ALGOL, Delphi, SNOBOL, Pascal, C, C++, Visual C++, Visual C #.NET, COBOL, BASIC, LISP, PROLOG, Visual Basic, Visual FoxPro, Java, and Ada.

For this reason, programs written in a high-level programming language may take longer to execute and use up more memory than programs written in an assembly language. High-level programming languages are more similar to normal human languages than low-level programming languages. Therefore, it is easier for the programmer to write larger and complicated programs faster.

.End.

参考译文 快速阅读1 编程语言介绍

如果没有程序给予计算机指令，计算机是无法运行的。专门编写程序的人被称为计算机程序员。他们用编程语言来构建程序。

在计算机科学中，编程语言是用来编写可被计算机运行的一系列指令的人工语言。这种语言与英语、汉语等自然语言不同，它是被称为代码的专门词汇、短语及符号的集合。人们用特定的且极为严格的方式把这些代码组合起来指导计算机运行。

无论如何，自然语言都不适合计算机编程，因为它们能引起歧义，也就是说它们的词汇和语法结构可以用多种方式进行解释。用于计算机编程的语言必须具有简单的逻辑结构，而且它们的语法、拼写和标点符号的规则必须精确。

编程语言几乎可以追溯到20世纪40年代数字计算机发明之时。自从第一代电子计算机在第二次世界大战中用于遥测计算以来，计算机语言已发生了巨大的变化。

早期程序员使用最原始的计算机指令——机器语言来工作。这些指令由一长串的0、1代码组成。最早的汇编语言，随着商业计算机的推出，出现于20世纪50年代末。汇编语言能将机器指令转换成易读、易管理的助记符，如ADD、MOV。

最早的过程语言是在20世纪50年代末到20世纪60年代初开发的。约翰·巴克斯创造了FORTRAN语言，之后格雷斯·霍珀创造了COBOL语言。第一种函数式语言是LISP，由约翰·麦卡锡编写于20世纪50年代末。这三种语言今天仍被广泛使用，但经历过大量修改。20世纪60年代末，出现了最早的面向对象语言，如SIMULA语言。在20世纪70年代，过程语言继续发展，出现了ALGOL、BASIC、PASCAL和C语言。

编程语言使用特定类型的语句或指令来给程序提供功能结构。程序中的一条语句表达一个意思简单的基本句子，其目的是给计算机提供一条基本指令。语句对允许的数据类型、数据如何被处理以及过程和函数的工作方式进行定义。

变量在程序中可以被赋予不同的值。变量具有的属性被称作类型，在许多编程语言中，一种关键的数据类型是指针。指针本身没有值；但是，它们含有计算机可以用来查找某个其他变量的信息——也就是说，它们指向另一个变量。

表达式是语句的一段，描述要对一些程序变量执行的一系列运算，如X+Y/Z，其中X、Y和Z为变量，运算方法为加和除。赋值语句给一个变量赋予来自某个表达式的值，而条件语句则是指定要被测试、然后选择接下来应该执行哪个语句的表达式。

过程与函数语句将某些代码块定义为以后可在程序中返回的进程或函数。这些语句也规定了程序员可以选择的变量与参数种类，以及当一个表达式使用过程或函数时代码将返回的值的类型。

最常用的编程语言具有很高的可移植性，可以用来有效地解决不同类型的计算问题。编程语言可被划分为低级编程语言和高级编程语言。

低级编程语言，比如汇编语言，是编程语言中最基础的类型。汇编语言非常接近机器语言，但仍然得翻译成机器语言。如何理解机器语言呢？机器语言是计算机能够直接识别的被称为比特的1、0代码的序列。

同样地，高级编程语言接近于人类自然语言，在计算机能够理解和处理之前也必须首先被翻译成机器语言。FORTRAN、ALGOL、Delphi、SNOBOL、Pascal、C、C++、Visual C++、Visual C#.NET、COBOL、BASIC、LISP、PROLOG、Visual Basic、Visual FoxPro、Java和Ada都是高级编程语言。

由于上述原因，与用汇编语言编写的程序比较起来，用高级编程语言编写的程序可能运行的时间更长，占用的内存更多。高级编程语言比低级编程语言更类似于正常的人类语言，因此，程序员容易更快地编写更庞大和更复杂的程序。

Fast Reading Two Safe Shopping Online

Fast Reading Two

Privacy of personal information is a significant issue for some consumers. Many consumers wish to avoid spam and telemarketing which could result from supplying contact information to an online merchant. In response, many merchants promise not to use consumer information for these purposes.

Many websites keep track of consumer shopping habits in order to suggest items and other websites to view. Brick-and-mortar stores also collect consumer information. Some ask for a shopper's address and phone number at checkout, though consumers may refuse to provide them. Many larger stores use the address information encoded on consumers' credit cards (often without their knowledge) to add them to a catalog mailing list. This information is obviously not accessible to the merchant when paying in cash.

Many online malls use cookies to track the user activity for showing relevant results to maximize the shopping experience. It can also be used to track your personal details. Thus before you pass your personal information, ensure the credibility of the online merchant. Good companies post their privacy policy (that is how they are going to use the personal information about you) on the website. Read carefully!

The important considerations after privacy are the security and safety features used by online malls. Remember that good websites are made in compliance with industrial standards such as SSL (secured socket layer). These standards use encryption technology to transfer information from your computer to online merchants' server.

By using SSL, the information you sent is scrambled. This means it is not possible to get details without encryption code. Since this is done automatically in merchants' server, it can be ensured that your personal details are secure. Thus make sure that you always do payment over SSL.

When the company you want to deal with is new to you, try to get maximum information about them before making any orders. Keep your password secret and make it in such a way that others may not be in a position to guess it.

Credit card transactions are considered to be the safest mode of payment for online shopping. Make yourself understand the company's policies, especially how they are going to keep your financial and personal data secured. Keep printed copies of purchase order and confirmation details, so that they can be used in the event of disputes.

.End.

参考译文 快速阅读2 安全的网上购物

对于一些消费者来说，个人信息保密是一个重要的问题。许多消费者都希望摆脱因向网上商家提供联系信息而出现的垃圾邮件和电话推销。作为回应，许多商家承诺不会出于这些目的来使用消费者信息。

许多网站会保留消费者的购物习惯，以便向其推荐其他商品或网站。实体店也会收集消费者信息，一些商家要求顾客在付款时提供地址和电话号码，但消费者可以拒绝提供。许多大商店会使用消费者信用卡上的地址信息编码(通常在他们不知情的情况下)，并将它们添加到邮寄目录中，这些信息显然不是向商家付现金时留下的。

许多网上商城使用cookie跟踪用户活动，以显示相关搜索结果来优化购物体验。它也可以用来记录你的个人信息，因此，在你留下个人信息之前，要确保网上商家的可信度。好的公司会在网站上发布其隐私政策(即它们将如何使用你的个人信息)，仔细阅读这些条款。

排在隐私之后的关于网上商城的主要考量方面是安全性和安全保护措施。记住，好的网站是符合工业标准如SSL(安全套接字层)的。这些标准使用加密技术将你电脑中的信息传输到在线商家的服务器上。

使用SSL发送信息，信息会经过置乱处理，这意味着没有加密代码是无法了解详细信息的。因为这是由商家服务器自动完成的，所以可以保证你的个人信息是安全的。因此，确保你总是通过SSL付款。

如果你要和新店家打交道，在发出任何订单前，一定要最大限度地获取它们的信息。保护好你的密码，确保别人不能轻易猜到你的密码。

网上购物中信用卡交易被认为是最安全的付款方式。要清楚交易公司的政策，尤其是在如何保证你的金融和个人信息安全方面的政策。保留打印的采购订单和细节凭证副本，以便在发生纠纷时使用。

Ex 1 What is online shopping? Try to describe it in a few words from the passage.

Ex 2 How many types are there in the E-Commerce? Describe them in a few words.

Ex 3 Fill in the table below by matching the corresponding Chinese or English equivalents.

	零售商
online	
	银行汇款
BTB	
	电子数据交换
E-Commerce	
	注册

Part B Practical Learning

Task One

Task Two

Training Target

In this part, there are two tasks in English environment. You should complete these tasks in groups under the joint guidance of professional teachers and laboratory teachers, so as to train and improve yours ability to complete professional tasks in English environment.

Task One Discuss the Differences Between Buyer and Seller

The important things for seller:

First: Online form should attract customers.

Secondly: The quality of the products should be good.

Thirdly: Ensure customer information security.

Fourthly: Right service attitude is irreplaceable.

The important things for buyer:

First: Find the product you want to buy.

When purchasing anything online, start by checking if the stores you usually shop have an online store. If you can't find a familiar name, don't hesitate to ask; you can find many helpful people online in chat rooms, forums, or mailing lists. You can ask for references or see if a site has any customer comments. You can also check the Better Business Bureau to see if there have been any complaints against a company.

Secondly: Read everything about the products carefully.

One of the most important things is to read all the information you can find about the specific product you are ordering. The last thing you want to do is receiving an item with a wrong size or a wrong color, etc. Do not rely solely on pictures provided. If the online store does not have enough information about the product, you can try to E-mail or call to see if they can clarify it for you.

Thirdly: Payment Options.

Check to see what kind of payment a site accepts. The most common way to buy things online is to use a credit card, which is very safe these days as long as a site has a secure connection. Some sites will let you pay by check. With this option, you usually place your order and then mail the payment, and the store will hold onto your order until the payment is received.

Fourthly: Shipping Cost.

One of the deciding factors for you when it comes to ordering craft supplies online is the shipping cost. In some cases, the shipping charge may be more than the cost of your purchase; again, read everything carefully! But sometimes people bite the bullet simply for the convenience of getting items shipped directly to their homes.

Last but not least: How about returns?

Make sure you are clear about the site's return policy. Some places have very strict policies and also charge restocking fees. You should be aware of what you are up against in case you need to return your products.

Task Two Trade Online

After finishing the online store design and knowing the principles of online trade, in this task, students will trade online. Some students should act as the sellers, and other students should act as the buyers. Then they can trade online.

Part C Occupation English

Occupation English

In this part, there is an English dialogue in real life and work environment. You will play the roles of A and B and read the dialogue aloud to practice your ability to use English.

Discussion on E-commerce
关于电子商务的讨论
Role setting: student (A), student (B)
角色设置：学生（A），学生（B）

微软技术支持人员要非常熟悉常用软件的操作，特别是视窗操作系统和办公系列软件。

A: What have you bought online recently? 你最近在网上买什么了？

B: Wow, I not only bought books, clothes, shoes, but also facial cleanser. My God, I can't count it! 哇，我不仅买了书、衣服、鞋子，还有洗面奶。我的天，数不过来了！

A: Ha ha, that is food and clothing, daily necessities and so on. They are all bought online. Actually, I'm about the same. 哈哈，就是吃穿日常用品等，都是从网上买的。其实，我也差不多。

B: Yes, now the online E-commerce is more convenient and fast. You can place an order and wait for the goods to be received. If offline, you still have to waste time to go to the store, and then choose, and you have to bring it back by yourself. It's really troublesome. 是呀，现在线上的电商更方便、快捷呀。下单等着收货就行了。要是线下，你还得浪费时间去店里，然后挑选，还要自己再拿回来，麻烦死了。

A: And now I don't worry about fakes. The quality is guaranteed. I love Jingdong shopping. How about you? 而且现在我也不太担心假货问题，质量都挺有保障的。我爱上京东买东西，你呢？

B: I will visit many stores, such as Taobao, Tmall, Jingdong, and sometimes Dangdang.我会逛好多店，比如淘宝，天猫，京东也去，有时也去当当买。

Word Building

前缀/后缀由一个或几个字母组成，放在词根或单词之前/之后，组成一个新词。

(1) auto-（前缀）：自动的

alarm 报警器 ———— autoalarm 自动报警器

code 编码 ———— autocode 自动编码

(2) -ee（后缀）：被……的人，受……的人

employ 雇用 ———— employee 雇员

test 测试 ———— testee 被测验者

(3) -th（后缀）：动作，性质，过程，状态

true 真实的，真正的 ———— truth 事实，真理

weal 福利，幸福 ———— wealth 财富

(4) -al（后缀）：属于……的，与……有关的

digit 数字 ———— digital 数字的

Ex Try your best to guess the meaning of each word on the right according to the clues given on the left.

train	训练（动词）	trainee	________
pay	薪水（名词）	payee	________
grow	生长（动词）	growth	________
deep	深的（形容词）	depth	________
education	教育（名词）	educational	________
nature	自然（名词）	natural	________
rotation	旋转（名词）	autoratation	________
biography	传记（名词）	autobiography	________

Exercises

Ex 1 What is E-Commerce? Try to describe it in a few words from the passage.

Ex 2 Fill in the table below by matching the corresponding Chinese or English equivalents.

	电子商务
BTB	
	网上购物
BTC	
	贝宝
restricting access	
	信用卡

Ex 3 Choose the best answer for each of the following statements according to the text we've learnt.

1. E-Commerce is doing business through __________.
   A. interfaces　　B. programs
   C. kernel techniques　　D. electronic media
2. Which is not the type of E-Commerce?
   A. BTB.　　B. BTC.
   C. ETS.　　D. Internet Explore.
3. Which is the widely accepted method of payment for online shopping?
   A. Credit card.
   B. E-checks.
   C. PayPal.
   D. Bank transfer.
4. Many online malls use __________ to track the user's activity for showing relevant results to maximize shopping experience.
   A. cookies
   B. Windows NT/2000
   C. Windows 98
   D. credibility
5. Good companies post their __________ on the website.
   A. privacy policy
   B. multi-user operating
   C. IE 6.0
   D. A and C

Project Nine

Order Tracking

Part A Theoretical Learning

Part B Practical Learning

Part C Occupation English

Part A Theoretical Learning

Training Target

In this part, our target is to improve the speed of reading professional articles and the comprehension ability of the reader. We have marked specialized key words and some flexible sentences. Try to grasp the main idea of each paragraph.

Skill One Internet of things

The Internet of Things (IoT) is a system of interrelated computing devices, mechanical and digital machines provided with unique identifiers (UIDs) and the ability to transfer data over a network without requiring human-to-human or human-to-computer interaction.

The definition of the Internet of things has evolved **due** to the **convergence** of multiple technologies, real-time analytics, machine learning, **commodity sensors**, and **embedded** systems. Traditional fields of embedded systems, wireless sensor networks, control systems, automation (including home and building automation), and others all contribute to enabling the Internet of things. In the consumer market, IoT technology is most synonymous with products pertaining to the concept of the "smart home", including devices and appliances (such as lighting fixtures, thermostats, home security systems and cameras, and other home appliances) that support one or more common **ecosystems**, and can be controlled via devices associated with that ecosystem, such as smartphones and smart speakers. **genomics**, **connectomics**, complex physics simulations, biology and environmental research.

History

The main concept of a network of smart devices was discussed as early as 1982, with a modified Coca-Cola vending machine at Carnegie Mellon University becoming the first Internet-connected appliance, able to report its inventory and whether newly loaded drinks were cold or not. Mark Weiser's 1991 paper on ubiquitous computing, "The Computer of the 21st Century", as well as academic venues such as UbiComp and PerCom produced the contemporary vision of the

the Internet of Things (IoT) 物联网
identifier [aɪˈdentɪfaɪə(r)] *n.* 标识符
unique identifier (UID) 唯一标识符
due [djuː] *adj.* 应得的 *n.* 某人应有的权益
convergence [kənˈvɜːdʒəns] *n.* 汇集
sensor [ˈsensə(r)] *n.* 传感器；探测设备
commodity [kəˈmɒdəti] *n.* 天然货物
embed [ɪmˈbed] *v.* 把……嵌入；埋置
embedded systems 嵌入式系统
ecosystem [ˈiːkəʊsɪstəm] *n.* 生态系统
genomics [dʒɪˈnəʊmɪks] *n.* 基因组学
connectomics 连接组学
momentum [məˈmentəm] *n.* 动量；动力

IoT. In 1994, Reza Raji described the concept in IEEE Spectrum as "[moving] small packets of data to a large set of nodes, so as to integrate and automate everything from home appliances to entire factories". Between 1993 and 1997, several companies proposed solutions like Microsoft's at Work or Novell's NEST. The field gained **momentum** when Bill Joy **envisioned** device-to-device communication as a part of his "Six Webs" framework, presented at the World Economic Forum at Davos in 1999.

The term "Internet of things" was **coined** by Kevin Ashton of Procter & Gamble, later MIT's Auto-ID Center, in 1999, though he prefers the phrase "Internet for things". At that point, he viewed **radio-frequency identification (RFID)** as essential to the Internet of things, which would allow computers to manage all individual things.

Defining the Internet of things as "simply the point in time when more 'things or objects' were connected to the Internet than people", Cisco Systems estimated that the IoT was "born" between 2008 and 2009, with the things/people ratio growing from 0.08 in 2003 to 1.84 in 2010.

The key driving force behind the Internet of things is the MOSFET (metal-oxide-semiconductor field-effect transistor, or MOS transistor),which was originally invented by Mohamed M. Atalla and Dawon Kahng at Bell Labs in 1959. The MOSFET is the basic building block of most modern electronics, including computers, smartphones, tablets and Internet services. MOSFET scaling **miniaturization** at a **pace predicted** by Dennard scaling and Moore's law has been the driving force behind technological advances in the electronics industry since the late 20th century. MOSFET scaling has been extended into the early 21st century with advances such as reducing power **consumption**, **silicon**-on-**insulator** (SOI) **semiconductor** device **fabrication**, and multi-core processor technology, leading up to the Internet of things, which is being driven by MOSFETs scaling down to **Nano electronic** levels with reducing energy consumption.

Architecture

IoT system architecture, in its simplistic view, consists of three **tiers**: Tier 1: Devices, Tier 2: the Edge Gateway, and Tier 3: the Cloud. Devices include networked things, such as the sensors and **actuators** found in IoT equipment, particularly those that use protocols such as Modbus, Bluetooth, Zigbee, or proprietary protocols, to connect to an Edge Gateway. The Edge Gateway consists of sensor data **aggregation**

envision [ɪnˈvɪʒn] *v.* 想象；设想

coin [kɔɪn] *n.* 硬币 *v.* 制造

radio-frequency identification（RFID） 射频识别

Skill One

miniaturization [ˌmɪnətʃəraɪ'zeɪʃn] *n.* 小型化；微型化

pace [peɪs] *n.* 步；步速 *v.* 恒速行走；踱步

predict [prɪˈdɪkt] *v.* 预言

consumption [kənˈsʌmpʃn] *n.* 耗尽；消耗

silicon [ˈsɪlɪkən] *n.* 硅

insulator [ˈɪnsjʊleɪtə(r)] *n.* 隔离物；绝缘体

semiconductor [ˌsemɪkənˈdʌktə(r)] *n.* 半导体

fabrication [ˌfæbrɪ'keɪʃn] *n.* 制造；建造

systems called Edge Gateways that provide functionality, such as pre-processing of the data, securing connectivity to cloud, using systems such as WebSockets, the event hub, and, even in some cases, edge analytics or fog computing. Edge Gateway layer is also required to give a common view of the devices to the upper layers to **facilitate** in easier management. The final tier includes the cloud application built for IoT using the microservices architecture, which are usually **polyglot** and **inherently** secure in nature using HTTPS/OAuth. It includes various database systems that store sensor data, such as time series databases or asset stores using backend data storage systems (e.g. Cassandra, PostgreSQL).The cloud tier in most cloud-based IoT system features event **queuing** and messaging system that handles communication that **transpires** in all tiers. Some experts classified the three-tiers in the IoT system as edge, platform, and enterprise and these are connected by proximity network, access network, and service network, **respectively**.

Building on the Internet of things, the web of things is an architecture for the application layer of the Internet of things looking at the convergence of data from IoT devices into Web applications to create innovative use-cases. In order to program and control the flow of information in the Internet of things, a predicted architectural direction is being called BPM Everywhere which is a blending of traditional process management with process mining and special capabilities to automate the control of large numbers of coordinated devices.

Network architecture

The Internet of things requires huge scalability in the network space to handle the surge of devices. IETF 6LoWPAN would be used to connect devices to IP networks. With billions of devices being added to the Internet space, IPv6 will play a major role in handling the network layer scalability. IETF's Constrained Application Protocol, ZeroMQ, and MQTT would provide lightweight data transport.

Fog computing is a viable alternative to prevent such large burst of data flow through Internet. The edge devices' computation power to analyses and process data is extremely limited. Limited processing power is a key attribute of IoT devices as their purpose is to supply data about physical objects while remaining autonomous. Heavy processing requirements use more battery power harming IoT's ability to operate. Scalability is easy because IoT devices simply supply data through the internet to a server with sufficient processing power.

.End.

Nano electronic 纳米电子学
tier [tɪə(r)]
n. 一排；一层；等级
actuator ['æktʃʊeɪtə(r)]
n. 激励者；促动器
aggregation [ˌægrɪ'geɪʃn]
n. 聚集；聚集体；聚合
facilitate [fəˈsɪlɪteɪt]
v. 使……更容易；使便利
polyglot [ˈpɒliglɒt]
adj. 多语的
inherently [ɪnˈherəntli]
adv. 天性地；固有地
queue [kjuː]
n. 行列 *v.* 排队
transpire [trænˈspaɪə(r)]
v. 被人知道，泄露
respectively [rɪˈspektɪvli]
adv. 各自地；分别地

Key Words

Identifier *n.* 标识符	unique identifier 唯一标识符
due *adj.* 应得的 *n.* 某人应有的权益	convergence *n.* 汇集
sensor *n.* 传感器；探测设备	commodity *n.* 天然货物
embed *v.* 把……嵌入；埋置	the Internet of Things 物联网
embedded systems 嵌入式系统	ecosystem *n.* 生态系统
genomics *n.* 基因组学	connectomics 连接组学
momentum *n.* 动量；动力	envision *v.* 想象，设想
coin *n.* 硬币 *v.* 制造	radio-frequency identification 射频识别
miniaturization *n.* 小型化；微型化	pace *n.* 步；步速 *v.* 恒速行走；踱步
predict *v.* 预言	consumption *n.* 耗尽；消耗
silicon *n.* 硅	insultor *n.* 隔离物；绝缘体
semiconductor *n.* 半导体	fabrication *n.* 制造；建造
Nano electronic 纳米电子学	tier *n.* 一排；一层；等级
actuator *n.* 激励者；促动器	aggregation *n.* 聚集；聚集体；聚合
facilitate *v.* 使……更容易；使便利	polyglot *adj.* 多语的
inherently *adv.* 天性地；固有地	queue *n.* 行列 *v.* 排队
transpire *v.* 被人知道，泄露	respectively *adv.* 各自地；分别地

参考译文 技能1 物联网

物联网（IoT）是一个由相互关联的计算设备、机械和数字机器组成的系统，这些设备具有唯一标识符（UID）和通过网络传输数据的能力，而不需要人与人或人与计算机交互。

由于多种技术、实时分析、机器学习、商品传感器和嵌入式技术的融合，物联网的定义不断演化。嵌入式系统、无线传感器网络、控制系统、自动化（包括家庭和楼宇自动化）等传统领域都有助于实现物联网。在消费市场中，物联网技术是与“智能家居”概念相关的产品的同义词，包括支持一个或多个常见生态系统的设备和电器（如照明设备、恒温器、家庭安全系统和摄像头以及其他家用电器），而且可以通过智能手机和智能音箱等与该生态系统相关的设备进行控制。基因组学，连接组学，复杂物理模拟，生物学和环境研究。

历史

智能设备网络的主要概念早在1982年就被讨论过，卡内基梅隆大学（Carnegie Mellon University）改装的可口可乐自动售货机（Coca-Cola）成为第一台联网设备，能够报告其库存和新装饮料是否凉。马克·韦瑟1991年关于普适计算的论文《21世纪的计算机》，以及UbiComp和PerCom等学术场所提出了物联网的当代构想。1994年，Reza Raji将IEEE Spectrum中的概念描述为“将小数据包移动到一个大的节点集，以便集成和自动化从家用电器到整个工厂的一切”。从1993年到1997年，有几家公司提出了微软的“在工作中”或“Novell的巢穴”

这样的解决方案。1999年在达沃斯举行的世界经济论坛上，比尔·乔伊设想将设备间通信作为其“六网”框架的一部分，这一领域获得了发展势头。

“物联网”一词是由凯文阿什顿在1999年首创的，他当时在保洁公司工作，后来在麻省理工学院担当自动识别中心工作。那时，他认为无线射频识别（RFID）是物联网的基本要素，它将允许计算机管理所有的个人物品。

思科系统公司将物联网定义为“仅仅是连接到互联网上的事物或对象多于人的时间点”，思科系统公司估计物联网是在2008年至2009年期间“诞生”的，物/人比率从2003年的0.08增长到2010年的1.84。

物联网背后的关键驱动力是MOSFET（金属氧化物半导体场效应晶体管，简称MOS晶体管），它最初由Mohamed M.Atalla和Dawon Kahng于1959年在贝尔实验室发明。MOSFET是大多数现代电子产品的基本组成部分，包括计算机、智能手机、平板电脑和互联网服务。自20世纪末以来，以Dennard scaling和Moore定律预测的速度进行的MOSFET小型化一直是推动电子工业技术进步的动力。随着诸如降低功耗、绝缘体上硅（SOI）半导体器件制造和多核处理器技术的进步，MOSFET的规模扩展到了21世纪初，导致了物联网，这是由MOSFET驱动的，通过降低能耗，缩小到纳米电子水平。

架构

物联网系统架构从简单的角度来看，由三层组成：第1层：设备，第2层：边缘网关，第3层：云。设备包括联网的东西，如IoT设备中的传感器和执行器，尤其是那些使用Modbus、Bluetooth、Zigbee或专有协议的设备，连接到边缘网关。边缘网关由称为边缘网关的传感器数据聚合系统组成，这些系统提供诸如数据预处理、保护与云的连接等功能，使用WebSocket、事件中心等系统，甚至在某些情况下，边缘分析或雾计算。边缘网关层还需要向上层提供设备的通用视图，以便于管理。最后一层包括使用microservices架构为IoT构建的云应用程序，microservices架构通常是多语言的，使用HTTPS/OAuth本质上是安全的。它包括存储传感器数据的各种数据库系统，例如使用后端数据存储系统的时间序列数据库或资产存储（如Cassandra，PostgreSQL）。大多数基于云的物联网系统的云层具有事件队列和消息传递系统，该系统处理所有层发生的通信。一些专家将IoT系统中的三层划分为边缘层、平台层和企业层，它们通过近距离网络、接入网和服务网络进行连接。

物联网是建立在物联网基础上的物联网应用层架构，着眼于物联网设备数据融合到网络应用中，以创造创新的用例。为了对物联网中的信息流进行编程和控制，一个预计的架构方向被称为“无处不在的BPM”，它将传统的流程管理与流程挖掘以及自动化控制大量协同设备的特殊功能相结合。

网络体系结构

物联网需要网络空间的巨大可扩展性来处理设备的激增。IETF 6LoWPAN将用于将设备连接到IP网络。随着数十亿台设备被添加到互联网空间，IPv6将在处理网络层可扩展性方面发挥重要作用。IETF的受限应用协议ZeroMQ和MQTT将提供轻量级数据传输。

雾计算是一个可行的替代方案，以防止这种大规模的数据流通过互联网。边缘设备的计算能力分析和处理数据是极其有限的。有限的处理能力是物联网设备的一个关键属性，因为它们的目的是在保持自主的同时提供有关物理对象的数据。繁重的处理要求使用更多的电池电量，损害物联网的运行能力。可扩展性很容易，因为物联网设备只需通过互联网向具有足够处理能力的服务器提供数据。

Skill Two 5G

In **telecommunications**, 5G is the fifth generation technology standard for **cellular** networks, which cellular phone companies began **deploying** worldwide in 2019, the planned successor to the 4G networks which provide connectivity to most current cellphones. Like its predecessors, 5G networks are cellular networks, in which the service area is divided into small geographical areas called **cells**. All 5G wireless devices in a cell are connected to the Internet and telephone network by radio waves through a local **antenna** in the cell. The main advantage of the new networks is that they will have greater **bandwidth**, giving higher download speeds, **eventually** up to 10 **gigabits** per second (Gbit/s). Due to the increased bandwidth, it is expected that the new networks will not just serve cellphones like existing cellular networks, but also be used as general Internet service providers for **laptops** and **desktop** computers, competing with existing **ISPs** such as **Cable Internet**, and also will make possible new applications in internet of things (IoT) and machine to machine areas. Current 4G cellphones will not be able to use the new networks, which will require new 5G enabled wireless devices.

The increased speed is achieved partly by using higher-frequency radio waves than current cellular networks. However, higher-frequency radio waves have a shorter range than the frequencies used by previous cell phone towers, requiring smaller cells. So to ensure wide service, 5G networks operate on up to three frequency bands, low, medium, and high. A 5G network will be composed of networks of up to 3 different types of cells, each requiring different antennas, each type giving a different **tradeoff** of download speed **VS** distance and service area. 5G cellphones and wireless devices will connect to the network through the highest speed antenna within range at their location.

Low-band 5G uses a similar frequency range as current 4G cellphones, 600~700 MHz, giving download speeds a little higher than 4G: 30-250 **megabits** per second (Mbit/s). Low-band cell towers will have a similar range and **coverage** area to current 4G towers. Mid-band 5G uses microwaves of 2.5~3.7 GHz, currently allowing speeds of 100-900 Mbit/s, with each cell tower providing service up to several miles in **radius**. This level of service is the most widely deployed, and should be available in most metropolitan areas in 2020. Some countries are not implementing low-band, making this the minimum service level. High-band 5G currently uses frequencies of 25~39

telecommunication [ˌtelɪkəˌmjuːnɪ'keɪʃən] *n.* 电信；电讯

cellular ['seljələ(r)] *adj.* （无线电话）蜂窝状的

deploy [dɪ'plɔɪ] *v.* 部署

cell [sel] *n.* 小房间；细胞

antenna [æn'tenə] *n.* 触角；天线

bandwidth ['bændwɪdθ] *n.* 带宽

eventually [ɪ'ventʃuəli] *adv.* 最后；终于

gigabit ['ɡɪɡəbɪt] *n.* 千兆比特

laptop ['læptɒp] *n.* 笔记本电脑

desktop ['desktɒp] *n.* 台式计算机

ISP 互联网服务供应商

Cable Internet 有线互联网

tradeoff ['treɪdɒf] *n.* 权衡

VS 对决

megabit ['meɡəbɪt] *n.* 百万比特

coverage ['kʌvərɪdʒ] *n.* 覆盖范围（或方式）

radius ['reɪdiəs] *n.* 半径（长度）；半径范围

GHz, near the bottom of the **millimeter** wave band, although higher frequencies may be used in the future. It often achieves download speeds of a gigabit per second (Gbit/s), **comparable** to cable internet. However, millimeter waves (mmWave or mmW) have a more limited range, requiring many small cells. They have trouble passing through some types of walls and windows. Due to their higher costs, current plans are to deploy these cells only in dense **urban** environments and areas where crowds of people **congregate** such as sports **stadiums** and **convention** centers. The above speeds are those achieved in actual tests in 2020, and speeds are expected to increase during **rollout**.

millimeter ['mɪlɪˌmiːtə(r)] *n.* 毫米
comparable ['kɒmpərəbl] *adj.* 类似的；可比较的
urban ['ɜːbən] *adj.* 城市的；都市的；城镇的
congregate ['kɒŋgrɪgeɪt] *v.* 群集；聚集；集合
convention [kən'venʃn] *n.* 大会
stadium ['steɪdiəm] *n.* 体育场
rollout ['rəʊˌlaʊt] *n.* 推出

Application areas

The **ITU-R** has defined three main application areas for the enhanced capabilities of 5G. They are **Enhanced Mobile Broadband (eMBB)**, **Ultra Reliable Low Latency Communications (URLLC)**, and **Massive Machine Type Communications (mMTC)**.Only eMBB is deployed in 2020; URLLC and mMTC are several years away in most locations.

ITU-R国际电信联盟
eMBB 大流量移动宽带业务
URLLC （无人驾驶、工业自动化等业务）超可靠低延迟通信
mMTC 大规模物联网业务

Enhanced Mobile Broadband (eMBB) uses 5G as a progression from 4G **LTE** mobile broadband services, with faster connections, higher throughput, and more capacity.

LTE(Long Term Evolution) 长期演进

Ultra-Reliable Low-Latency Communications (URLLC) refer to using the network for mission **critical** applications that require uninterrupted and robust data exchange.

critical ['krɪtɪkl] *adj.* 关键的

Massive Machine-Type Communications (mMTC) would be used to connect to a large number of devices, 5G technology will connect some of the 50 billion connected to IoT devices. Most will use the less expensive Wi-Fi. **Drones**, transmitting via 4G or 5G, will **aid** in **disaster** recovery efforts, providing real-time data for emergency **responders**. Most cars will have a 4G or 5G cellular connection for many services. Autonomous cars do not require 5G, as they have to be able to operate where they do not have a network connection.While remote surgeries have been performed over 5G, most remote **surgery** will be performed in **facilities** with a fiber connection, usually faster and more reliable than any wireless connection.

drone [drəʊn] *n.* 无人机
aid [eɪd] *n.* 援助 *v.* 帮助
disaster [dɪ'zɑːstə(r)] *n.* 灾难
responder [rɪs'pɒndə(r)] *n.* 应答器
surgery ['sɜːdʒəri] *n.* 外科手术
facility [fə'sɪləti] *n.* 设施

Performance

Speed

5G speeds will range from 50 Mbit/s to over a gigabit.The fastest 5G, known as mmWave. As of July 3, 2019, mmWave had a top speed of 1.8 Gbit/s on AT&T's 5G network.

Skill Two

Sub-6 GHz 5G (mid-band 5G), by far the most common, will usually deliver between 100 and 400 Mbit/s, but will have a much farther reach than mmWave, especially outdoors.

Low-band **spectrum** offers the farthest area coverage but is slower than the others.

5G NR speed in sub-6 GHz bands can be slightly higher than the 4G with a similar amount of spectrum and antennas,although some 3GPP 5G networks will be slower than some advanced 4G networks, such as T-Mobile's LTE/LAA network, which achieves 500+ Mbit/s in Manhattan and Chicago.The 5G specification allows LAA (**License Assisted Access**) as well, but LAA in 5G has not yet been demonstrated. Adding LAA to an existing 4G **configuration** can add hundreds of megabits per second to the speed, but this is an extension of 4G, not a new part of the 5G standard.

The similarity in terms of **throughput** between 4G and 5G in the existing bands is because 4G already approaches the Shannon limit on data communication rates. 5G speeds in the less common millimeter wave spectrum, with its much more **abundant** bandwidth and shorter range, and **hence** greater frequency reuseability, can be **substantially** higher.

Latency

In 5G, the "air latency" in equipment shipping in 2019 is 8~12 milliseconds. The latency to the server must be added to the "air latency" for most comparisons. Verizon reports the latency on its 5G early deployment is 30 ms:Edge Servers close to the towers can reduce latency to 10~20 ms; 1~4 ms will be extremely rare for years outside the lab.

spectrum ['spektrəm] *n.* 谱；光谱

License Assisted Access (LAA) 许可频谱辅助接入

configuration [kən,fɪgə'reɪʃn] *n.* 布局

throughput ['θruːpʊt] *n.* 吞吐量

abundant [ə'bʌndənt] *adj.* 大量的

hence [hens] *adv.* 因此

substantially [səb'stænʃəli] *adv.* 基本上

latency ['leɪtənsi] *n.* 延迟

.End.

Key Words

telecommunication *n.* 电信；电讯	deploy *v.* 部署
cellular *adj.* （无线电话）蜂窝状的	cell *n.* 小房间；细胞
antenna *n.* 触角；天线	bandwidth *n.* 带宽
eventually *adv.* 最后；终于	gigabit *n.* 千兆比特
laptop *n.* 笔记本电脑	desktop *n.* 台式计算机
tradeoff *n.* 权衡	VS 对决
megabit *n.* 百万比特	coverage *n.* 覆盖范围（或方式）
radius *n.* 半径（长度）；半径范围	millimeter *n.* 毫米

comapareble *adj.* 类似的；可比较的

urban *adj.* 城市的；都市的；城镇的

congregate *v.* 群集；聚集；集合

critical *adj.* 关键的

aid *n.* 援助 *v.* 帮助

responder *n.* 应答器

facility *n.* 设施

configuration *n.* 布局

abundant *adj.* 大量的

substantially *adv.* 基本上

ISP 互联网服务供应商

ITU-R 国际电信联盟

URLLC（无人驾驶、工业自动化等业务）超可靠低延迟通信

mMTC 大规模物联网业务

License Assisted Access 许可频谱辅助接入

convention *n.* 大会

stadium *n.* 体育场

rollout *n.* 推出

drone *n.* 无人机

disaster *n.* 灾难

surgery *n.* 外科手术

spectrum *n.* 谱；光谱

throughput *n.* 吞吐量

hence *adv.* 因此

latency *n.* 延迟

Cable Internet 有线互联网

eMBB 大流量移动宽带业务

LTE 长期演进

参考译文 技能2 5G技术

在电信领域，5G是蜂窝网络的第五代技术标准，移动电话公司于2019年开始在全球范围内部署该标准，计划成为4G网络的继承者。4G网络为目前大多数手机提供连接服务。与其前身一样，5G网络是蜂窝网络，其中的服务区被划分为小的地理区域，称为“蜂窝”。小区内所有5G无线设备借助小区内的本地天线通过无线电波连接到互联网和电话网。新网络的主要优势在于，它们将拥有更高的带宽，提供更高的下载速度，最终将达到每秒10千兆比特（Gbit/s）。由于带宽的增加，预计新网络将不像现有的蜂窝网络那样只为手机服务，但也将作为笔记本电脑和台式电脑的通用互联网服务提供商，与有线互联网等现有互联网服务提供商竞争，也将使物联网（IoT）和机器对机器领域的新应用成为可能。目前的4G手机将无法使用新的网络，这将需要新的5G无线设备。

速度的提高部分是通过使用比当前蜂窝网络更高频率的无线电波来实现的。然而，高频无线电波的射程比以前的手机发射塔使用的频率范围小，适合更小的小区。因此，为了确保广泛的服务，5G网络最多可在低、中、高三个频段上运行。5G网络将由多达3种不同类型的小区组成，每种小区需要不同的天线，每种类型在下载速度与距离和服务面积之间有着不同的权衡。5G手机和无线设备将通过其所在位置范围内的最高速度天线连接到网络：

低频段5G使用与当前4G手机相似的频率范围，600～700兆赫，下载速度略高于4G:30～250兆位每秒（Mbit/s）。低频段蜂窝塔的范围和覆盖区域将与当前的4G发射塔相似。中频5G使用2.5～3.7 GHz的微波，目前允许的速度为100～900 Mbit/s，每个蜂窝塔可提供半径达数千米的服务。这一级别的服务是部署最广泛的，到2020年将在大多数大都市地区提供。一些国家没有实施低频段，这是最低服务水平。目前，高频5G使用的频率为25～39 GHz，接近毫米波波段的底部，尽管未来可能会使用更高频率。它的下载速度通常达到每秒千兆位（Gbit/s），与有线互联网相当。然而，毫米波（mmWave或mmW）的范围更为有限，需要许多小单元。他们很难穿过某些类型的墙和窗户。由于成本较高，目前的计划是只在密集的城市环境和人群聚集的地区（如体育场和会议中心）部署这些单元。上述速度是在2020年实际测试中达到的速度，预计在推出过程中速度会提高。

应用领域

ITU-R定义了增强5G能力的三个主要应用领域。它们是增强型移动宽带（eMBB）、超可靠低延迟通信（URLLC）和大规模机器型通信（mMTC）。只有eMBB在2020年才部署；URLLC和mMTC在大多数地区还需要几年的时间。

增强型移动宽带（eMBB）将5G作为4G-LTE移动宽带服务的一个进步，具有更快的连接、更高的吞吐量和更大的容量。

超可靠低延迟通信（urlc）是指将网络用于需要不间断和健壮的数据交换的关键应用程序。

大规模机器类型通信（mMTC）将用于连接大量设备，5G技术将连接500亿个已连接物联网设备中的一些。大多数将使用较便宜的Wi-Fi。无人机通过4G或5G传输，将有助于灾后恢复工作，为应急响应人员提供实时数据。大多数汽车都将有4G或5G蜂窝连接，用于多种服务。自动驾驶汽车不需要5G，因为它们必须能够在没有网络连接的地方进行操作。虽然远程手术是通过5G进行的，但大多数远程手术将在具有光纤连接的设施中进行，通常比任何无线连接都要快、更可靠。

性能

速度

5G的速度将从大约50 Mbit/s到超过千兆比特。最快的5G，称为mmWave。截至2019年7月3日，mmWave在AT&T的5G网络上的最高速度为1.8 Gbit/s。

到目前为止，最常见的5G（中频5G）频率通常在100～400 Mbit/s，但比mmWave的覆盖范围更广，尤其是在户外低频段频谱提供最远的区域覆盖，但比其他频段慢。

5G NR在低于6GHz频段的速度可能略高于频谱和天线数量相同的4G网络，尽管一些3GPP 5G网络将比一些先进的4G网络（如在曼哈顿和芝加哥的T-Mobile的LTE/LAA网络，可达500+Mbit/s）慢，在曼哈顿和芝加哥5G规范也允许LAA（许可证辅助访问），但5G中的LAA尚未得到验证。在现有的4G配置中添加LAA可以使速度提高数百兆位/秒，但这是4G的扩展，而不是5G标准的新组成部分。

在现有频段中，4G和5G在吞吐量方面的相似性是因为4G在数据通信速率上已经接近香农极限。在不太常见的毫米波频谱中，5G的速度可以大大提高，因为它具有更丰富的带宽和更短的距离，因此具有更高的频率可重用性。

延迟

在5G中，2019年设备发货时的“空气延迟”为8～12毫秒。大多数比较时，到服务器的延迟必须添加到“空气延迟”中。Verizon报告称，其5G早期部署的延迟为30毫秒：靠近发射塔的边缘服务器可以将延迟减少到10～20毫秒；1～4毫秒在实验室之外的几年里是极为罕见的。

Fast Reading One Application of Internet of Things

Consumer applications

A growing portion of IoT devices are created for consumer use, including connected vehicles, home automation, wearable technology, connected health, and appliances with remote monitoring capabilities.

Fast Reading One

Smart home

IoT devices are a part of the larger concept of home automation, which can include lighting, heating and air conditioning, media and security systems and camera systems. Long-term benefits could include energy savings by automatically ensuring lights and electronics are turned off.

A smart home or automated home could be based on a platform or hubs that control smart devices and appliances. For instance, using Apple's HomeKit, manufacturers can have their home products and accessories controlled by an application in iOS devices such as the iPhone and the Apple Watch. This could be a dedicated app or iOS native applications such as Siri. This can be demonstrated in the case of Lenovo's Smart Home Essentials, which is a line of smart home devices that are controlled through Apple's Home app or Siri without the need for a Wi-Fi bridge. There are also dedicated smart home hubs that are offered as standalone platforms to connect different smart home products and these include the Amazon Echo, Google Home, Apple's HomePod, and Samsung's SmartThings Hub. In addition to the commercial systems, there are many non-proprietary, open source ecosystems; including Home Assistant, OpenHAB and Domoticz.

Elder care

One key application of a smart home is to provide assistance for those with disabilities and elderly individuals. These home systems use assistive technology to accommodate an owner's specific disabilities. Voice control can assist users with sight and mobility limitations while alert systems can be connected directly to cochlear implants worn by hearing-impaired users. They can also be equipped with additional safety features. These features can include sensors that monitor for medical emergencies such as falls or seizures. Smart home technology applied in this way can provide users with more freedom and a higher quality of life.

Organizational applications

Medical and healthcare

The Internet of Medical Things (IoMT) is an application of the IoT for medical and health related purposes, data collection and analysis for research, and monitoring. The IoMT has been referred to as "Smart Healthcare", as the technology for creating a digitized healthcare system, connecting available medical resources and healthcare services.

IoT devices can be used to enable remote health monitoring and emergency notification systems. These health monitoring devices can range from blood pressure and heart rate monitors to advanced devices capable of monitoring specialized implants, such as pacemakers, Fitbit electronic wristbands, or advanced hearing aids. Some hospitals have begun implementing "smart beds" that can detect when they are occupied and when a patient is attempting to get up. It can also adjust itself to ensure appropriate pressure and support is applied to the patient without the manual interaction of nurses. Advances in plastic and fabric electronics fabrication methods have enabled ultra-low cost, use-and-throw IoMT sensors. These sensors, along with the required RFID electronics, can be fabricated on paper or e-textiles for wireless powered disposable sensing devices. Applications have been established for point-of-care medical diagnostics, where portability and low system-complexity are essential.

The application of the IoT in healthcare plays a fundamental role in managing chronic diseases and in disease prevention and control. Remote monitoring is made possible through the connection of powerful wireless solutions. The connectivity enables health practitioners to capture patient's data and applying complex algorithms in health data analysis.

Transportation

The IoT can assist in the integration of communications, control, and information processing across various transportation systems. Application of the IoT extends to all aspects of transportation systems (i.e. the vehicle, the infrastructure, and the driver or user). Dynamic interaction between these components of a transport system enables inter- and intra-vehicular communication, smart traffic control, smart parking, electronic toll collection systems, logistics and fleet management, vehicle control, safety, and road assistance. In Logistics and Fleet Management, for example, an IoT platform can continuously monitor the location and conditions of cargo and assets via wireless sensors and send specific alerts when management exceptions occur (delays, damages, thefts, etc.). This can only be possible with IoT technology and its seamless connectivity among devices. Sensors such as GPS, Humidity, and Temperature send data to the IoT platform and then the data is analyzed and then sent to the users. This way, users can track the real-time status of vehicles and can make appropriate decisions. If combined with Machine Learning, then it also helps in reducing traffic accidents by introducing drowsiness alerts to drivers and providing self-driven cars too.

V2X communications

In vehicular communication systems, vehicle-to-everything communication (V2X), consists of three main components: vehicle to vehicle communication (V2V), vehicle to infrastructure communication (V2I) and vehicle to pedestrian communications (V2P). V2X is the first step to autonomous driving and connected road infrastructure.

Building and home automation

IoT devices can be used to monitor and control the mechanical, electrical and electronic systems used in various types of buildings (e.g., public and private, industrial, institutions, or residential) in home automation and building automation systems.

The IoT can realize the seamless integration of various manufacturing devices equipped with sensing, identification, processing, communication, actuation, and networking capabilities. Based on such a highly integrated smart cyber-physical space, it opens the door to create whole new business and market opportunities for manufacturing. Network control and management of manufacturing equipment, asset and situation management, or manufacturing process control bring the IoT within the realm of industrial applications and smart manufacturing as well. The IoT intelligent systems enable rapid manufacturing of new products, dynamic response to product demands, and real-time optimisation of manufacturing production and supply chain networks, by networking machinery, sensors and control systems together.

Infrastructure applications

Monitoring and controlling operations of sustainable urban and rural infrastructures like bridges, railway tracks and on- and offshore wind-farms is a key application of the IoT. The IoT infrastructure can be used for monitoring any events or changes in structural conditions that can compromise safety and increase risk. The IoT can benefit the construction industry by cost saving, time reduction, better quality workday, paperless workflow and increase in productivity. It can help in taking faster decisions and save money with Real-Time Data Analytics. It can also be used for scheduling repair and maintenance activities in an efficient manner, by coordinating tasks between different service providers and users of these facilities. IoT devices can also be used to control critical infrastructure like bridges to provide access to ships. Usage of IoT devices for monitoring and operating infrastructure is likely to improve incident management and emergency response coordination, and quality of service, up-times and reduce costs of operation in all infrastructure related areas. Even areas such as waste management can benefit from automation and optimization that could be brought in by the IoT.

Military applications

The Internet of Military Things (IoMT) is the application of IoT technologies in the military domain for the purposes of reconnaissance, surveillance, and other combat-related objectives. It is heavily influenced by the future prospects of warfare in an urban environment and involves the use of sensors, munitions, vehicles, robots, human-wearable biometrics, and other smart technology that is relevant on the battlefield.

.End.

参考译文 快速阅读1 物联网的应用

越来越多的物联网设备是为消费者使用而设计的，包括联网车辆、家庭自动化、可穿戴技术、联网健康和具有远程监控功能的设备。

智能家居

物联网设备是家庭自动化更大概念的一部分，包括照明、供暖和空调、媒体和安全系统以及摄像头系统。其长期效益在于通过自动关闭电灯和电子设备来节约能源。

智能家庭或自动化家庭可以基于控制智能设备和电器的平台或集线器。例如，使用苹果的HomeKit，制造商可以通过iPhone和Apple等iOS设备中的应用程序来控制他们的家用产品和配件。这个可能是专门应用程序或iOS原生应用程序，如Siri。这可以从联想的智能家居软件中得到证明，这是一系列智能家居设备，通过苹果的家庭应用程序或Siri进行控制，而不需要Wi-Fi桥牌。那里它们也是专门的智能家居集线器，作为独立平台提供，用于连接不同的智能家居产品，包括Amazon Echo、Google home、苹果HomePod和三星的SmartThings Hub。除了商业系统之外，还有许多非专有的、开源的生态系统；包括Home Assistant、OpenHAB和Domoticz。

老年护理

智能家居的一个关键应用是为残疾人和老年人提供帮助。这些家庭系统使用辅助技术来适应业主的特殊残疾。听力受损的用户可直接佩戴耳蜗，以帮助视力受损的用户。它们还可以配备额外的安全功能。这些功能可以包括传感器，监测医疗紧急情况，如跌倒或癫痫。以这种方式应用的智能家居技术可以为用户提供更多的自由和更高的生活质量。

医疗和保健

医疗物联网是物联网在医疗和健康相关领域的应用，用于研究和监测的数据收集和分析。以及IoMT被称为“智能医疗”，作为创建数字化医疗系统的技术，连接可用的医疗资源和医疗服务。

物联网设备可用于实现远程健康监测和紧急通知系统。这些健康监测设备包括血压和心率监测仪以及能够监测专业植入物的先进设备，如起搏器、Fitbit电子腕带或高级助听器。一些医院已经开始实施“智能床”，可以检测到他们何时被占用，以及病人何时尝试起来。它还可以调整自身，确保在不需要护士人工交互的情况下，适当地向患者施加压力和支持。塑料和织物电子制造方法的进步使得超低成本、使用和丢弃IoMT传感器成为可能。这些传感器以及所需的RFID电子设备，可在纸上或电子纺织品上制造，用于无线动力一次性传感设备.应用已经建立了用于医疗诊断的护理点，在那里，便携性和低系统复杂性是必不可少的。

物联网在医疗保健中的应用，对慢性病的管理和疾病的防治具有重要意义。通过连接强大的无线解决方案，远程监控成为可能。这种连接使卫生从业人员能够捕获患者的数据，并在健康数据分析中应用复杂算法。

V2X通信

在车辆通信系统中，车辆到一切通信（V2X）由三个主要部分组成：车辆到车辆通信（V2V）、车辆到基础设施通信（V2I）和车辆到行人通信（V2P）。V2X是实现自动驾驶和连接道路基础设施的第一步。

楼宇和家庭自动化

物联网设备可用于监测和控制家庭自动化和楼宇自动化系统中各种类型建筑物（例如公共和私人、工业、机构或住宅）中使用的机械、电气和电子系统。

建筑和家庭自动化

物联网设备可用于监测和控制家庭自动化和楼宇自动化系统中各种类型建筑物（例如公共和私人、工业、机构或住宅）中使用的机械、电气和电子系统。

物联网可以实现各种制造设备的无缝集成，这些设备配备传感、识别、处理、通信、驱动和网络功能。基于如此高度集成的智能网络物理空间，它为制造业创造全新的业务和市场机遇打开了大门。制造设备、资产和状况管理或制造过程控制的网络控制和管理将物联网纳入工业应用和智能制造领域好吧。那物联网智能系统可快速制造新产品、动态响应产品需求和实时优化制造生产和供应链网络，通过网络化的机械、传感器和控制系统在一起。

基础设施应用程序

监测和控制可持续城市和农村基础设施（如桥梁、铁路轨道和陆上和海上风电场）的运行是物联网物联网基础设施可用于监测可能危及安全和增加风险的任何事件或结构条件变化。物联网可以通过节省成本、缩短时间、提高工作日质量、实现无纸化工作流程和提高生产率，使建筑业受益。它可以帮助您更快地做出决策，并通过实时数据分析节省资金。它还可以通过协调不同服务提供商和用户之间的任务，有效地安排维修和维护活动设施. IoT设备还可用于控制关键基础设施，如桥梁，以提供对船舶的访问。使用物联网设备监控和运营基础设施可能会改善事件管理和应急响应协调，提高服务质量、正常运行时间并降低所有基础设施相关领域的运营成本。甚至像废物管理这样的领域也可以从物联网带来的自动化和优化中获益。

军事应用

军事物联网（IoT）是物联网技术在军事领域的应用，用于侦察、监视和其他与作战有关的目标。它在很大程度上受城市环境下未来战争前景的影响，涉及传感器、弹药、车辆、机器人、人类可穿戴生物特征识别和其他与战场相关的智能技术的使用。

Fast Reading Two　Enabling Technologies for LoT

There are many technologies that enable the IoT crucial to the field is the network used to communicate between devices of an IoT installation, a role that several wireless or wired technologies may fulfill:

Fast Reading Two

Addressability

The original idea of the Auto-ID Center is based on RFID-tags and distinct identification through the Electronic Product Code. This has evolved into objects having an IP address or URI. An alternative view, from the world of the Semantic Web focuses instead on making all things (not just those electronic, smart, or RFID-enabled) addressable by the existing naming protocols, such as URI. The objects themselves do not converse, but they may now be referred to by other agents, such as powerful centralised servers acting for their human owners. Integration with the Internet implies that devices will use an IP address as a distinct identifier. Due to the limited address space of IPv4 (which allows for 4.3 billion different addresses), objects in the IoT will have to use the next generation of the Internet protocol (IPv6) to scale to the extremely large address space required. Internet-of-things devices additionally will benefit from the stateless address auto-configuration present in IPv6, as it reduces the configuration overhead on the hosts, and the IETF 6LoWPAN header compression. To a large extent, the future of the Internet of things will not be possible without the support of IPv6; and consequently, the global adoption of IPv6 in the coming years will be critical for the successful development of the IoT in the future.

Application layer

ADRC defines an application layer protocol and supporting framework for implementing IoT applications.

Short-range wireless

•Bluetooth mesh networking – Specification providing a mesh networking variant to Bluetooth low energy (BLE) with increased number of nodes and standardized application layer (Models).

•Light-Fidelity (Li-Fi) – Wireless communication technology similar to the Wi-Fi standard, but using visible light communication for increased bandwidth.

•Near-field communication (NFC) – Communication protocols enabling two electronic devices to communicate within a 4 cm range.

•Radio-frequency identification (RFID) – Technology using electromagnetic fields to read data stored in tags embedded in other items.

•Wi-Fi – Technology for local area networking based on the IEEE 802.11 standard, where devices may communicate through a shared access point or directly between individual devices.

•ZigBee – Communication protocols for personal area networking based on the IEEE 802.15.4 standard, providing low power consumption, low data rate, low cost, and high throughput.

•Z-Wave – Wireless communications protocol used primarily for home automation and security applications

Medium-range wireless

•LTE-Advanced – High-speed communication specification for mobile networks. Provides enhancements to the LTE standard with extended coverage, higher throughput, and lower latency.

•5G – 5G wireless networks can be used to achieve the high communication requirements of the IoT and connect a large number of IoT devices, even when they are on the move.

Long-range wireless

•Low-power wide-area networking (LPWAN) – Wireless networks designed to allow long-range communication at a low data rate, reducing power and cost for transmission. Available LPWAN technologies and protocols: LoRaWan, Sigfox, NB-IoT, Weightless, RPMA.

•Very small aperture terminal (VSAT) – Satellite communication technology using small dish antennas for narrowband and broadband data.

Wired

•Ethernet – General purpose networking standard using twisted pair and fiber optic links in conjunction with hubs or switches.

•Power-line communication (PLC) – Communication technology using electrical wiring to carry power and data. Specifications such as HomePlug or G.hn utilize PLC for networking IoT devices.

参考译文 | 快速阅读2 物联网支持技术

多项技术的支持使物联网成为可能。该领域的关键是物联网设备之间通信的网络，而各种有线或无线技术使之得以实现。

可寻址能力

自动识别中心的最初想法是基于射频识别标签和通过电子产品代码进行清晰识别。这已经演变成具有IP地址或URI的对象。从语义网的角度来看，另一种观点是专注于让所有事物(不仅仅是那些电子的、智能的或支持射频识别的东西)都可以通过现有的命名协议来寻址，比如

URI。这些对象本身并不交流，但它们现在可能会被其他代理所引用，比如为它们的人类主人服务的强大的集中式服务器。与互联网的集成意味着设备将使用一个IP地址作为独特的标识符。由于IPv4的地址空间有限(允许43亿个不同的地址)，物联网中的对象将不得不使用下一代互联网协议(IPv6)来扩展到所需的超大地址空间。物联网设备还将受益于IPv6中存在的无状态地址自动配置，因为它减少了主机上的配置开销，以及IETF 6LoWPAN报头压缩。很大程度上，没有IPv6的支持，物联网的未来是不可能的；因此，IPv6在未来几年的全球采用对于物联网未来的成功发展至关重要。

应用层

ADRC为实现物联网应用定义了应用层协议和支持框架。

短程无线

蓝牙网状网络-规范提供了蓝牙低能耗(BLE)的网状网络变体，增加了节点数量和标准化应用层(模型)。

光保真(Li-Fi)——类似于Wi-Fi标准的无线通信技术，但使用可见光通信来增加带宽。

近场通信——通信协议使两个电子设备能够在4厘米范围内通信。

射频识别(RFID)——利用电磁场读取存储在嵌入其他物品的标签中的数据的技术。

无线网络——基于IEEE 802.11标准的局域网技术，设备可以通过共享接入点或直接在单个设备之间进行通信。

紫蜂——基于IEEE 802.15.4标准的个人区域网络通信协议，提供低功耗、低数据速率、低成本和高吞吐量。

震波——主要用于家庭自动化和安全应用的无线通信协议。

中程无线

长期演进——高级——移动网络的高速通信规范。通过扩展的覆盖范围、更高的吞吐量和更低的延迟，增强了LTE标准。

5G —— 5G无线网络可用于实现物联网的高通信要求，并连接大量物联网设备，即使它们在移动中。

远程无线

低功耗广域网(LPWAN)——无线网络旨在以低数据速率进行远程通信，降低传输功耗和成本。可用的LPWAN技术和协议:LoRaWan、Sigfox、NB-IoT、失重、RPMA。

甚小孔径终端(VSAT)——卫星通信技术，使用小型碟形天线传输窄带和宽带数据。

有线的

以太网——通用网络标准，结合集线器或交换机使用双绞线和光纤链路。

电力线通信——使用电线传输电力和数据的通信技术。HomePlug或G.hn等规范利用可编程逻辑控制器对物联网设备进行联网。

Ex 1 What is the IoT?Try to give a brief summary according to the passage.

Ex 2 Fill in the table below by matching the corresponding Chinese or English equivalents.

IoT	
	传感器
laptop	
	台式机
semiconductor	
	射频识别

Ex 3 Choose the best answer to the following statements according to the text we've learnt.

1. The definition of the ____________ has evolved due to the convergence of multiple technologies, real-time analytics, machine learning, commodity sensors, and embedded systems.
   A. Internet of things　　B. programs
   C. kernel techniques　　D. account numbers
2. ____________ system architecture, in its simplistic view, consists of three tiers: Tier 1: Devices, Tier 2: the Edge Gateway, and Tier 3: the Cloud.
   A. Platform　　B. IoT
   C. Graphical User Interface　　D. Senor
3. ____________ is a viable alternative to prevent such large burst of data flow through Internet.
   A. Fog computing　　B. Cloud computing
   C. Windows 10　　D. RFID
4. The key driving force behind the Internet of things is ____________ (metal-oxide-semiconductor field-effect transistor, or MOS transistor),which was originally invented by Mohamed M. Atalla and Dawon Kahng at Bell Labs in 1959
   A. platform　　B. Plug and Play
   C. Media Player　　D. the MOSFET
5. The term "Internet of things" was coined by Kevin Ashton of Procter & Gamble, later MIT's Auto-ID Center, in 1999, though he prefers the phrase "____________".
   A. Internet for things　　B. IoT
   C. Linux　　D. 3ds Max

Part B Practical learning

Task One

Task Two

Training Target

In this part, there are two tasks in English environment. You should complete these tasks in groups under the joint guidance of professional teachers and laboratory teachers, so as to train and improve your ability to complete professional tasks in English environment.

In this part, the students must finish two special tasks in English environment, Under the guidance of the specialized English teacher and the teacher teaching the network query in the computer laboratory. The students must work with their companion in the same group.

Task One Students Complete an Online Shopping Transaction

The first task is for students to buy the things they need online, such as books, clothes, daily necessities, etc.

There are some information about the online shopping transaction. These information can help the student finish the task.

The students' needs can be purchased online in Jingdong or Dangdang. (Pic 9.1, Pic 9.2)

Pic 9.1 Jingdong

Pic 9.2 Dangdang

After entering the website, students can choose the goods they need.(Pic9.3,Pic9.4)

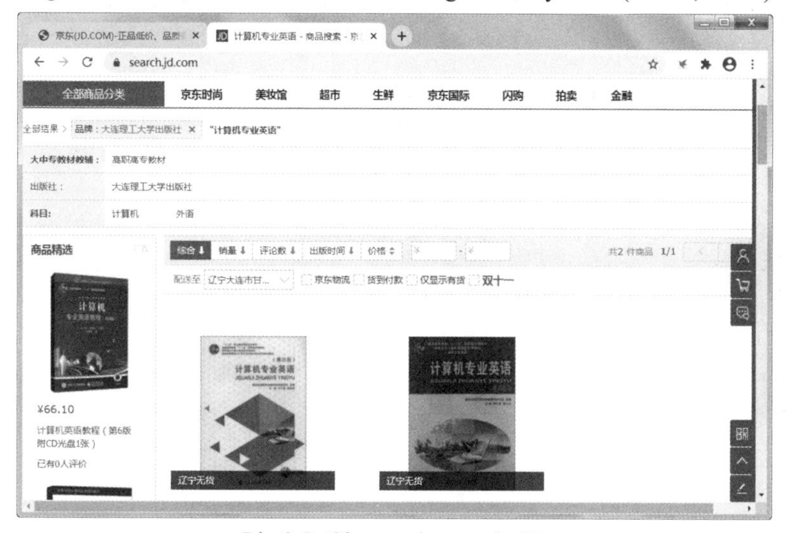

Pic 9.3 Choose the goods (1)

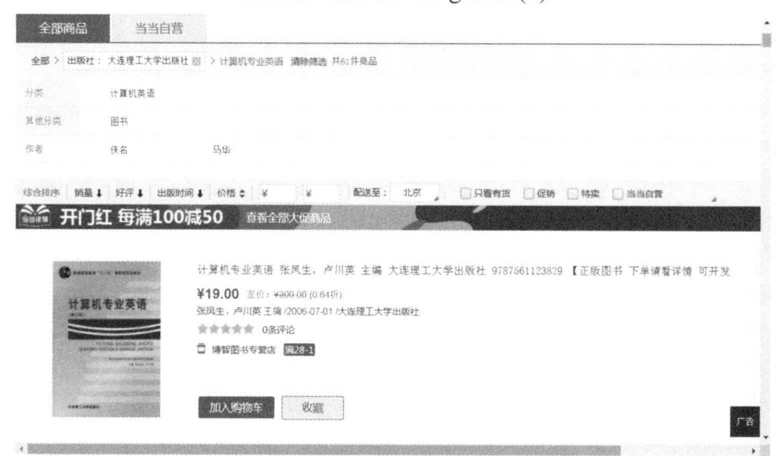

Pic 9.4 Choose the goods (2)

Task Two Order tracking

In this task, the students make a thorough inquiry on the commodity orders completed in task 1 and master the order dynamics.

Students can also query in express delivery, input the express order number to query the logistics information of the order. (such as Pic 9.5)

Pic 9.5 The order query system

After inputting the order number, the order query system will display the complete information of the order and the logistics situation. (Pic 9.6)

Pic 9.6 Display the complete information

Part C Occupation English

Occupation English

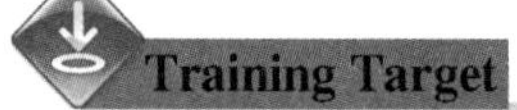

In this part, there is an English dialogue in real life and work environment. You will play the roles of A and B and read the dialogue aloud to practice your ability to use English.

Interview

面试

Role setting: interviewers of an enterprise (A), students of a school (B)

角色设置：公司面试官（A），学生（B）

A: Hello, nice to meet you. We are Huawei Shenzhen Branch.你好，很高兴认识你。我们是华为深圳分公司。

B: Hello, nice to meet you. I specially submitted a resume of Huawei. Huawei is a place where I admire and yearn for a job. Thank you for giving me this interview opportunity. I will strive for it.您好，很高兴认识你。我特意投了华为一份简历，华为是我敬仰和向往工作的地方，谢谢您给我这个面试的机会，我会努力争取。

A: Please introduce yourself first.那请你先自我介绍一下。

B: My name is Wang Hua. I graduated from Huazhong University of Science and Technology. I majored in computer communication technology and majored in network security.我叫王华，毕业于华中科技大学，计算机通信技术专业，还辅修了网络安全专业。

A: What did you learn in University?那谈谈你在大学学到了什么？

B: In school, I not only learned professional knowledge, but also learned how to cooperate with others, be friendly, hard-working and persevere by participating in some communities and student union activities. I also learned to bravely try the unknown field, and I also learned to stick to my dream.在学校，我不仅学到了专业知识，通过参加一些社团和学生会的活动让我学会如何与人团结协作、友善、努力、坚持。我还学会了对于未知的领域勇敢地去尝试，也学会了坚持自己的梦想。

A: Yes, young people should bravely try to pursue their own goals. Besides these, have you ever participated in voluntary activities?是呀，年轻人就该勇敢地去尝试，去不断的追求自己心中的目标。除了这些，你参加过义务活动吗？

B: Yes, I have been a volunteer, and I have been to a nursing home with young volunteers to take care of the elderly.参加过，我做过义工，还和青年志愿者一起去过敬老院，照顾老人等。

A: Good. Talk about your own greatest strength.很好，说说你自己最大的优点。

B: I am an honest, kind, responsible and patriotic person. And I'm also a person who does everything with all my strength. Once I set a goal, I will use all my strength and try my best! Even if I failed, I would not regret it. I'm not afraid of failure. Because I believe that as long as I persist, I will succeed. 我是一个正直、善良，有责任感和爱国之心得一个人。并且，我还是一个做事全力以赴的人。一旦确定了一个目标，我会用上自己全部的力量，并且很认真的尽力！就算我失败了也不后悔了。我不怕失败，因为我相信只要坚持就一定会成功。

A: Do you have any relevant work experience?你有相关的工作经验吗？

B: I did a part-time job in Baidu in the process of going to university, engaged in related work.我在上大学过程中在百度做过兼职，从事过相关工作。

A: Well, that's my question. Go back and wait for our notice. Good luck!好了，我的问题就到这，回去等我们的通知，祝你好运！

B: Thank you! Hope to give me this opportunity to enter the enterprise I admire and strive for it!谢谢！希望能给我这个机会，进入我仰慕的企业，为之奋斗！

Word Building

前缀/后缀由一个或几个字母组成，放在词根或单词之前/之后，组成一个新词。

（1）im-（前缀）表示“不，无，非”

possible 可能的 ———— impossible 不可能的

moral 道德的 ———— immoral 不道德的

（2）im-（前缀）表示“向内，进入”

prison 监狱 ———— imprison 监禁

pel 推 ———— impel 驱动

（3）a-(前缀)表示“不，无”

centric 中心的 ———— acentric 无中心的

social 好社交的 ———— asocial 不好社交的

（4）a-(前缀)表示“在…，…的”

sleep 睡觉(动词) ———— asleep 睡着的

side 旁边(名词) ———— aside 在边上

(5)名词后加-y，名词变成形容词

luck 幸运(名词) ———— lucky 幸运的

cloud 云(名词) ———— cloudy 多云的

Ex Translate the following words and try your best to guess the meaning of each word on the right according to the clues given on the left.

polite	礼貌的（形容词）	impolite________
partial	有偏见的（形容词）	impartial________
peril	危险(名词)	imperil________
mortal	不能永生的（形容词）	immortal________
pulse	跳动(动词)	impulse________
moral	道德的（形容词）	amoral________
political	政治的（形容词）	apolitical________
head	头(名词)	ahead________
live	活（动词）	alive________
wind	风(名词)	windy________
rain	雨(名词)	rainy________

Ex 1 What is 5G? Try to give a brief summary of this passage.

Ex 2 What is IoT? Try to give a brief summary of this passage.

Ex 3 What is dely? Try to give a brief summary of this passage.

Ex 4 Fill in the table below by matching the corresponding Chinese or English equivalents.

cellular	
	电信
bandwidth	
	国际电信联盟
mMTC	
	互联网服务供应商
insulator	
	大流量移动宽带业务
RFID	
	唯一标识符

Ex 5 Choose the best answer to for each of following statements according to the text we've learnt.

1. In telecommunications, ____________ is the fifth generation technology standard for cellular networks.

   A. 5G　　B. 2G

   C. 4G　　D. 3G

2. Like its predecessors, 5G networks are ____________, in which the service area is divided into small geographical areas called cells.

   A. internet

   B. cellular networks

   C. wireless network

   D. network

3. Current ____________ cellphones will not be able to use the new networks, which will require new 5G enabled wireless devices.

   A. 4G

   B. 5G

   C. 3G

   D.2G

4. The main concept of a network of smart devices was discussed as early as ____________ .

   A. 1982

   B. 1983

   C. 1986

   D. 1985

5. In 5G, the "____________" in equipment shipping in 2019 is 8~12 milliseconds.

   A. air latency

   B. delay

   C. fog computing

   D.A and B

Project Ten

Artificial Intelligence

Part A Theoretical Learning

Part B Practical Learning

Part C Occupation English

Part A Theoretical Learning

Training Target

In this part, our target is to improve the speed of reading professional articles and the comprehension ability of the reader. We have marked specialized key words and some flexible sentences. Try to grasp the main idea of each paragraph.

Skill One

Skill One | A short introduction of artificial intelligence

Artificial intelligence (AI), sometimes called machine intelligence, is intelligence demonstrated by machines, unlike the natural intelligence displayed by humans and animals. Leading AI textbooks define the field as the study of "**intelligent agents**": any device that **perceives** its environment and takes actions that maximize its chance of successfully achieving its goals. **Colloquially**, the term "artificial intelligence" is often used to describe machines (or computers) that **mimic** "**cognitive**" functions that humans associate with the human mind, such as "learning" and "problem solving".

As machines become increasingly capable, tasks considered to require "intelligence" are often removed from the definition of AI, a phenomenon known as the AI effect. A **quip** in Tesler's Theorem says "AI is whatever hasn't been done yet." For instance, optical character recognition is frequently excluded from things considered to be AI, having become a **routine** technology. Modern machine capabilities generally classified as AI include successfully understanding human speech, competing at the highest level in **strategic** game systems (such as chess and Go), autonomously operating cars, intelligent routing in content **delivery** networks, and military **simulations**.

Artificial intelligence was founded as an **academic discipline** in 1955, and in the years since has experienced several waves of optimism, followed by **disappointment** and the loss of funding (known as an "AI winter"), followed by new approaches, success and renewed funding. For most of its history, AI research has been divided into sub-fields that often fail to communicate with each other. These sub-fields are based on technical considerations, such as particular goals (e.g.

artificial [ˌɑ:tɪ'fɪʃl] *adj.* 人工的；人造的
intelligence [ɪn'telɪdʒəns] *n.* 智力
artificial intelligence (AI) 人工智能
demonstrate ['demənstreɪt] *v.* 证明；证实
intelligent agent 智慧代理人
perceive [pə'si:v] *v.* 注意到
colloquial [kə'ləʊkwiəl] *adj.* 口语化；口语的，通俗的
mimic ['mɪmɪk] *v.* 模仿
cognitive ['kɒgnətɪv] *adj.* 认知的
quip [kwɪp] *n.* 俏皮话；妙语
routine [ru:'ti:n] *n.* 常规
strategic [strə'ti:dʒɪk] *adj.* 战略性的
delivery [dɪ'lɪvəri] *n.* 交付
simulation [ˌsɪmju'leɪʃn] *n.* 模拟；仿真
academic [ˌækə'demɪk] *adj.* 学术的

"robotics" or "machine learning"), the use of particular tools ("logic" or artificial neural networks), or deep **philosophical** differences. Subfields have also been based on social factors (particular institutions or the work of particular researchers).

The traditional problems (or goals) of AI research include reasoning, knowledge representation, planning, learning, natural language processing, perception and the ability to move and manipulate objects. General intelligence is among the field's long-term goals. Approaches include statistical methods, computational intelligence, and traditional symbolic AI. Many tools are used in AI, including versions of search and mathematical **optimization**, artificial neural networks, and methods based on statistics, **probability** and economics. The AI field draws upon computer science, information engineering, mathematics, **psychology**, **linguistics**, philosophy, and many other fields.

Basics

Computer science defines AI research as the study of "intelligent agents": any device that perceives its environment and takes actions that maximize its chance of successfully achieving its goals. A more **elaborate** definition characterizes AI as "a system's ability to correctly interpret external data, to learn from such data, and to use those learnings to achieve specific goals and tasks through flexible adaptation.

A typical AI analyzes its environment and takes actions that maximize its chance of success. An AI's intended utility function (or goal) can be simple ("1 if the AI wins a game of Go, 0 otherwise") or complex ("Perform actions mathematically similar to ones that succeeded in the past"). Goals can be explicitly defined or induced. If the AI is programmed for "**reinforcement** learning", goals can be **implicitly** induced by **rewarding** some types of behavior or punishing others.

AI often revolves around the use of algorithms. An algorithm is a set of **unambiguous** instructions that a mechanical computer can execute. A complex algorithm is often built on top of other simpler algorithms.

Applications

AI is **relevant** to any intellectual task. Modern artificial intelligence techniques are **pervasive** and are too numerous to list here. Frequently, when a technique reaches mainstream use, it is no longer considered artificial intelligence; this phenomenon is described as the AI effect.

discipline ['dɪsəplɪn] *n.* 学科
disappointment [ˌdɪsə'pɔɪntmənt] *n.* 失望
philosophical [ˌfɪlə'sɒfɪkl] *adj.* 哲学的
reason ['riːzn] *v.* 推理
perception [pə'sepʃn] *n.* 感知
optimization [ˌɒptɪmaɪ'zeɪʃən] *n.* 最佳（优）化；优选法
probability [ˌprɒbə'bɪləti] *n.* 概率
psychology [saɪ'kɒlədʒi] *n.* 心理学
linguistic [lɪŋ'gwɪstɪk] *n.* 语言学
elaborate [ɪ'læbərət] *adj.* 精心制作的 *v.* 详尽阐述
reinforcement [ˌriːɪn'fɔːsmənt] *n.* 加强
implicitly [ɪm'plɪsɪtli] *adv.* 暗中地；含蓄地
reward [rɪ'wɔːd] *n.* 奖励
unambiguous [ˌʌnæm'bɪgjuəs] *adj.* 意思清楚的；明确的
relevant ['reləvənt] *adv.* 紧密相关的；切题的
pervasive [pə'veɪsɪv] *adj.* 遍布的

High-profile examples of AI include autonomous vehicles (such as drones and self-driving cars), medical **diagnosis**, creating art (such as poetry), proving mathematical theorems, playing games (such as Chess or Go), search engines (such as Google search), online assistants (such as Siri), image recognition in photographs, **spam** filtering, **predicting flight delays**, prediction of **judicial** decisions, **targeting** online **advertisements**, and **energy** storage.

In the twenty-first century, AI techniques have experienced a **resurgence** following concurrent advances in computer power, large amounts of data, and theoretical understanding; and AI techniques have become an essential part of the technology industry, helping to solve many challenging problems in computer science, software engineering and **operations research**.

.End.

diagnosis [ˌdaɪəg'nəʊsɪs] *n.* 诊断
spam [spæm] *n.* 垃圾邮件
predict [prɪ'dɪkt] *v.* 预告，预报
flight [flaɪt] *n.* 航程
delay [dɪ'leɪ] *n.* 延迟
judicial [dʒu'dɪʃl] *adj.* 法庭的
target ['tɑːgɪt] *v.* 把……作为攻击目标
advertisement [əd'vɜːtɪsmənt] *n.* 广告；启事
energy ['enədʒi] *n.* 能源
resurgence [rɪ'sɜːdʒəns] *n.* 复苏
operations research 运筹学

artificial *adj.* 人工的；人造的
artificial intelligence 人工智能
intelligent agent 智慧代理人
colloquial *adj.* 口语化；口语的，通俗的
cognitive *adj.* 认知的
routine *n.* 常规的
delivery *n.* 交付
academic *adj.* 学术的
disappointment *n.* 失望
reason *v.* 推理
optimization *n.* 最佳（优）化；优选法
psychology *n.* 心理学
elaborate *adj.* 精心制作的 *v.* 详尽阐述
implicitly *adv.* 暗中地；含蓄地
unambiguous *adj.* 意思清楚的；明确的
pervasive *adj.* 遍布的
spam *n.* 垃圾邮件
flight *n.* 航程
judicial *adj.* 法庭的
advertisement *n.* 广告；启事
resurgence *n.* 复苏
intelligence *n.* 智力
demonstrate *v.* 证明；证实
perceive *v.* 注意到
mimic *v.* 模仿
quip *n.* 俏皮话，妙语
strategic *adj.* 战略性的
simulation *n.* 模拟；仿真
discipline *n.* 学科
philosophical *adj.* 哲学的
perception *n.* 感知
probability *n.* 概率
linguistic *n.* 语言学
reinforcement *n.* 加强
reward *n.* 奖励
relevant *adj.* 紧密相关的；切题的
diagnosis *n.* 诊断
predict *v.* 预告，预报
delay *n.* 延迟
target *v.* 把……作为攻击目标
energy *n.* 能源
operations research 运筹学

参考译文 | 技能1 人工智能简介

人工智能（AI），有时称为机器智能，是由机器展示的智能，不同于人类和动物展示的自然智能。领先的人工智能教科书将这一领域定义为“智能代理”的研究：任何能够感知环境并采取行动以最大限度地提高成功实现目标的机会的设备。通俗地说，“人工智能”一词通常用来描述模仿“认知”功能的机器（或计算机），这些功能与人类思维相关，比如“学习”和“解决问题”。

随着机器的能力越来越强，被认为需要“智能”的任务往往会从人工智能的定义中消失，这种现象被称为人工智能效应。泰斯勒定理中有句俏皮话说，“人工智能是任何尚未实现的东西”。例如，光学字符识别常常被排除在人工智能之外，已经成为一项常规技术。通常被归为人工智能的现代机器能力包括成功理解人类语言，在战略游戏系统（如国际象棋和围棋）中处于最高水平的竞争对手，自动驾驶操作汽车，内容交付网络中的智能路由，以及军事模拟。

人工智能作为一门学科创立于1955年，此后几年经历了几波乐观主义浪潮，随后是失望和资金损失（被称为“人工智能冬天”），接着是新方法、成功和重新获得资金。在人工智能研究的大部分历史中，它一直被划分为多个子领域，而这些子领域往往无法相互交流。这些子领域基于技术考虑，例如特定目标（例如“机器人学”或“机器学习”）、特定工具（“逻辑”或“人工神经网络”）的使用，或深刻的哲学差异。子领域也基于社会因素（特定机构或特定研究人员的工作）。

人工智能研究的传统问题（或目标）包括推理、知识表示、计划、学习、自然语言处理、感知以及移动和操纵对象的能力。一般智能是该领域的长期目标之一。方法包括统计方法、计算智能，以及传统的象征性人工智能。人工智能中使用了许多工具，包括搜索和数学优化版本、人工神经网络以及基于统计学、概率论和经济学的方法。人工智能领域综合了计算机科学、信息工程、数学、心理学、语言学、哲学和许多其他领域。

基础

计算机科学将人工智能研究定义为对“智能代理”的研究：任何感知环境并采取行动以最大限度地提高成功实现其目标的机会的设备。更详细的定义将人工智能描述为“一个系统能够正确解释外部数据，从这些数据中学习，并通过灵活的适应，利用这些学习来实现特定的目标和任务。”

一个典型的人工智能会分析它的环境，并采取行动使其成功的机会最大化。一个AI的预期效用函数（或目标）可以是简单的（“如果AI赢了围棋游戏，则为1”，否则为0）或复杂（“在数学上执行与过去成功相似的动作”）。目标可以明确定义或归纳。如果人工智能被编程为“强化学习”，目标可以通过奖励某些类型的行为或惩罚其他类型的行为而隐含地诱导。

人工智能通常围绕着算法的使用。算法是一组机械计算机可以执行的明确指令。一个复杂的算法通常是建立在其他更简单的算法之上的。

应用

人工智能与任何智力任务有关。现代人工智能技术无处不在，不胜枚举。通常，当一项技术达到主流用途时，它就不再被认为是人工智能；这种现象被称为人工智能效应。

人工智能引人注目的例子包括自动驾驶汽车（如无人机和自动驾驶汽车）、医疗诊断、创作艺术（如诗歌）、证明数学定理、玩游戏（如象棋或围棋）、搜索引擎（如谷歌搜索）、在线助手（如Siri）、照片中的图像识别、垃圾邮件过滤、预测航班延误、司法判决预测、针对在线广告和能量储存。

在21世纪，随着计算机能力、大量数据和理论理解的同步发展，人工智能技术经历了一次复苏；人工智能技术已经成为科技行业的重要组成部分，帮助解决计算机科学、软件工程和运筹学中的许多具有挑战性的问题。

Skill Two The Applications of Artificial Intelligence

AI is relevant to any **intellectual** task. Modern artificial intelligence techniques are pervasive and are too numerous to list here.

Competitions and prizes

There are a number of competitions and prizes to promote research in artificial intelligence. The main areas promoted are: general machine intelligence, **conversational** behavior, **data-mining**, robotic cars, robot soccer and games.

Healthcare

Artificial intelligence is breaking into the healthcare industry by assisting doctors. According to Bloomberg Technology, Microsoft has developed AI to help doctors find the right **treatments** for cancer. There is a great amount of research and **drugs** developed relating to cancer. In detail, there are more than 800 medicines and **vaccines** to treat cancer. This negatively affects the doctors, because there are way too many options to choose from, making it more difficult to choose the right drugs for the patients. Microsoft is working on a project to develop a machine called "Hanover". Its goal is to memorize all the papers necessary to cancer and help predict which combinations of drugs will be most effective for each patient. One project that is being worked on at the moment is fighting **myeloid leukemia**, a **fatal** cancer where the treatment has not improved in **decades**. Another study was reported to have found that artificial intelligence was as good as trained doctors in **identifying** skin cancers. Another study is using artificial intelligence to try and monitor multiple high-risk patients, and this is done by asking each patient numerous questions based on data **acquired** from live doctor to patient interactions.

Automotive industry

Advancements in AI have contributed to the growth of the automotive industry through the creation and evolution of self-driving vehicles. As of 2016, there are over 30 companies **utilizing** AI into the creation of driverless cars.

intellectual [ˌɪntə'lektʃuəl] *adj.* 智力的
profile ['prəʊfaɪl] *n.* 简介，印象
conversational [ˌkɒnvə'seɪʃənl] *adj.* 口语的；交谈的；会话的
mining ['maɪnɪŋ] *n.* 采矿
data-mining 数据挖掘
healthcare ['helθkeə(r)] *n.* 医疗卫生；保健
treatment ['triːtmənt] *n.* 治疗
drug [drʌgz] *n.* 药
vaccine ['væksiːn] *n.* 疫苗
myeloid ['maɪəlɔɪd] *adj.* 骨髓的
leukemia [luː'kɪːmɪə] *n.* 白血病
fatal ['feɪtl] *adj.* 致命的
decade ['dekeɪd] *n.* 十年
identify [aɪ'dentɪfaɪ] *v.* 确认
acquire [ə'kwaɪə(r)] *v.* 获得
automotive [ˌɔːtə'məʊtɪv] *n.* 汽车
utilize ['juːtəlaɪz] v. 使用

Many components contribute to the functioning of self-driving cars. These vehicles incorporate systems such as **braking**, lane changing, **collision prevention**, navigation and **mapping**. Together, these systems, as well as high performance computers are integrated into one complex vehicle.

One main factor that influences the ability for a driver-less car to function is mapping. In general, the vehicle would be pre-programmed with a map of the area being driven. This map would include data on the **approximations** of street light and **curb** heights in order for the vehicle to be aware of its **surroundings**. However, Google has been working on an algorithm with the purpose of eliminating the need for pre-programmed maps and instead, creating a device that would be able to **adjust** to a variety of new surroundings. Some self-driving cars are not equipped with **steering wheels** or brakes, so there has also been research focused on creating an algorithm that is capable of maintaining a safe environment for the passengers in the vehicle through awareness of speed and driving conditions.

brake ['breɪk] v. 刹(车)
collision [kə'lɪʒn] n. 碰撞
prevention [prɪ'venʃn] n. 预防
map ['mæp] v. 绘制……的地图；了解信息
approximation [əˌprɒksɪ'meɪʃn] n. 近似值；粗略计算
curb [kɜːb] v. 控制
surrounding [sə'raʊndɪŋ] n. 环境
adjust [ə'dʒʌst] v. 调整
steering ['stɪərɪŋ] n. (车辆等的)转向装置
wheel [wiːl] n. 轮

Finance

Financial institutions have long used artificial neural network systems to detect **charges** or **claims** outside of the **norm**, **flagging** these for human **investigation**.

Use of AI in banking can be tracked back to 1987 when Security Pacific National Bank in USA set-up a Fraud Prevention Task force to counter the unauthorized use of debit cards. Apps like Kasisito and Moneystream are using AI in financial services

Banks use artificial intelligence systems to organize operations, maintain book-keeping, **invest** in **stocks**, and manage properties. AI can react to changes overnight or when business is not taking place.

AI has also reduced fraud and crime by monitoring behavioral patterns of users for any changes or anomalies.

charge ['tʃɑːdʒ] n. 要价，收费
claim [kleɪm] n. 索赔
norm [nɔːm] n. 标准
flag ['flæg] v. 标示
investigation [ɪnˌvestɪ'geɪʃn] n. 侦查；科学研究
invest [ɪn'vest] v. 投资
stock [stɒk] n. 股票

Video games

Artificial intelligence is used to generate intelligent behaviors primarily in non-player characters (NPCs), often simulating human-like intelligence.

With the development of other technologies, artificial intelligence technology is more and more mature, and its application in life will be more and more extensive.

.End.

Skill Two

Key Words

intellectual *adj.* 智力的	profile *n.* 简介，印象
conversational *adj.* 口语的；交谈的；会话的	
mining *n.* 采矿	data-mining 数据挖掘
healthcare *n.* 医疗卫生；保健	treatment *n.* 治疗
drug *n.* 药	vaccine *n.* 疫苗
myeloid *adj.* 骨髓的	leukemia *n.* 白血病
fatal *adj.* 致命的	decade *n.* 十年
identify *v.* 确认	acquire *v.* 获得
automotive *n.* 汽车	utilize *v.* 使用
brake *v.* 刹（车）	collision *n.* 碰撞
prevention *n.* 预防	map *v.* 绘制……的地图；了解信息
approximation *n.* 近似值；粗略计算	corb *v.* 控制
surrounding *n.* 环境	adjust *v.* 调整
steering *n.* （车辆等的）转向设置	charge *n.* 要价，收费
chaim *n.* 索赔	norm *n.* 标准
flag *v.* 标示	investigation *n.* 侦查；科学研究
invest *v.* 投资	stock *n.* 股票

参考译文 技能2 人工智能的应用

人工智能与任何智力任务都相关。现代人工智能技术无处不在，不胜枚举。

竞赛和奖励

有许多竞赛和奖项来推广人工智能的研究。推广的主要领域有：通用机器智能、会话行为、数据挖掘、机器人汽车、机器人足球和游戏。

医疗

人工智能正通过协助医生进入医疗保健行业。据彭博科技报道，微软已经开发出人工智能来帮助医生找到正确的癌症治疗方法。有大量与癌症有关的研究和药物开发。具体来说，治疗癌症的药物和疫苗有800多种。这对医生产生了负面影响，因为有太多的选择可供选择，使得为病人选择合适的药物变得更加困难。微软正在进行一个项目，开发一种叫作“汉诺威”的机器。它的目标是记忆所有与癌症有关的论文，并帮助预测哪些药物组合对每个患者最有效。目前正在进行的一个项目是对抗髓系白血病，这是一种致命的癌症，治疗方法几十年来一直没有改善。据报道，另一项研究发现，人工智能在识别皮肤癌方面与训练有素的医生不相上下。另一项研究是使用人工智能来监测多个高危患者，这是通过询问每个患者大量的问题来完成的，这些问题是基于实时医患互动获得的数据。

汽车工业

人工智能的进步促进了自动驾驶汽车的发明和发展，进而为促进了汽车行业的增长做出了贡献。截至2016年，有超过30家公司利用人工智能技术制造无人驾驶汽车。

许多部件都有助于自动驾驶汽车的功能。这些车辆包括刹车、变道、防撞、导航和地图绘制等系统。这些系统和高性能计算机一起集成到一个复杂的车辆中。

影响无驾驶员汽车运行能力的一个主要因素是地图。一般情况下，车辆会预先编程一张驾驶区域的地图。该地图将包括路灯和路缘高度的近似值数据，以便车辆了解周围环境。然而，谷歌一直在研究一种算法，其目的是消除对预先编程地图的需求，而是创建一种能够适应各种新环境的设备。一些自动驾驶汽车没有配备方向盘或刹车，因此也有研究集中在创建算法上通过对车速和驾驶条件的了解，为车内乘客维持一个安全的环境。

金融领域

金融机构长期以来一直使用人工神经网络系统来检测超出标准的收费或索赔，并将其标记，供人类调查。

人工智能在银行业的应用可以追溯到1987年，当时美国太平洋国家银行（Security Pacific National Bank）成立了一个防止欺诈的特别小组，以对付未经授权使用借记卡的行为。Kassito和Moneystream等应用程序正在金融服务中使用人工智能。

银行使用人工智能系统来组织业务、维持簿记、投资股票和管理财产。当业务在一夜之间发生变化时，人工智能可以做出反应。

人工智能还通过监控用户行为模式的任何变化或异常来减少欺诈和犯罪。

视频游戏

人工智能主要用于生成非玩家角色(NPCs)的智能行为，通常模拟类人智能。

随着其他技术的发展，人工智能技术越来越成熟，在生活中的应用也将越来越广泛

Fast Reading One The Approaches of Artificial Intelligence

Fast Reading One

There is no established unifying theory or paradigm that guides AI research. Researchers disagree about many issues. A few of the most long standing questions that have remained unanswered are these: should artificial intelligence simulate natural intelligence by studying psychology or neurology? Or is human biology as irrelevant to AI research as bird biology is to aeronautical engineering? Can intelligent behavior be described using simple, elegant principles (such as logic or optimization)? Or does it necessarily require solving a large number of completely unrelated problems? Can intelligence be reproduced using high-level symbols, similar to words and ideas? Or does it require “sub-symbolic” processing? John Haugeland, who coined the term GOFAI (Good Old-Fashioned Artificial Intelligence), also proposed that AI should more properly be referred to as synthetic intelligence, a term which has since been adopted by some non-GOFAI researchers.

Stuart Shapiro divides AI research into three approaches, which he calls computational psychology, computational philosophy, and computer science. Computational psychology is used to make computer programs that mimic human behavior. Computational philosophy, is used to develop an adaptive, free-flowing computer mind. Implementing computer science serves the goal of creating computers that can perform tasks that only people could previously accomplish. Together, the humanesque behavior, mind, and actions make up artificial intelligence.

Cybernetics and brain simulation

In the 1940s and 1950s, a number of researchers explored the connection between neurology, information theory, and cybernetics. Some of them built machines that used electronic networks to exhibit rudimentary intelligence, such as W. Grey Walter's turtles and the Johns Hopkins Beast. Many of these researchers gathered for meetings of the Teleological Society at Princeton University and the Ratio Club in England. By 1960, this approach was largely abandoned, although its elements would be revived in the 1980s.

Symbolic

When access to digital computers became possible in the middle 1950s, AI research began to explore the possibility that human intelligence could be reduced to symbol manipulation. The research was centered in three institutions: Carnegie Mellon University, Stanford and MIT, and each one developed its own style of research. John Haugeland named these approaches to AI "good old fashioned AI" or "GOFAI". During the 1960s, symbolic approaches had achieved great success at simulating high-level thinking in small demonstration programs. Approaches based on cybernetics or neural networks were abandoned or pushed into the background. Researchers in the 1960s and the 1970s were convinced that symbolic approaches would eventually succeed in creating a machine with artificial general intelligence and considered this the goal of their field.

Cognitive simulation

Economist Herbert Simon and Allen Newell studied human problem-solving skills and attempted to formalize them, and their work laid the foundations of the field of artificial intelligence, as well as cognitive science, operations research and management science. Their research team used the results of psychological experiments to develop programs that simulated the techniques that people used to solve problems. This tradition, centered at Carnegie Mellon University would eventually culminate in the development of the Soar architecture in the middle 1980s.

Logic-based

Unlike Newell and Simon, John McCarthy felt that machines did not need to simulate human thought, but should instead try to find the essence of abstract reasoning and problem solving, regardless of whether people used the same algorithms. His laboratory at Stanford (SAIL) focused on using formal logic to solve a wide variety of problems, including knowledge representation, planning and learning. Logic was also the focus of the work at the University of Edinburgh and elsewhere in Europe which led to the development of the programming language Prolog and the science of logic programming.

Anti-logic or scruffy

Researchers at MIT (such as Marvin Minsky and Seymour Papert) found that solving difficult problems in vision and natural language processing required ad-hoc solutions — they argued that there was no simple and general principle (like logic) that would capture all the aspects of intelligent behavior. Roger Schank described their "anti-logic" approaches as "scruffy" (as opposed to the "neat" paradigms at CMU and Stanford). Commonsense knowledge bases (such as Doug Lenat's Cyc) are an example of "scruffy" AI, since they must be built by hand, one complicated concept at a time.

Knowledge-based

When computers with large memories became available around 1970, researchers from all three traditions began to build knowledge into AI applications. This "knowledge revolution" led to the development and deployment of expert systems (introduced by Edward Feigenbaum), the first truly successful form of AI software. The knowledge revolution was also driven by the realization that enormous amounts of knowledge would be required by many simple AI applications.

Sub-symbolic

By the 1980s progress in symbolic AI seemed to stall and many believed that symbolic systems would never be able to imitate all the processes of human cognition, especially perception, robotics, learning and pattern recognition. A number of researchers began to look into "sub-symbolic" approaches to specific AI problems. Sub-symbolic methods manage to approach intelligence without specific representations of knowledge.

Embodied intelligence

This includes embodied, situated, behavior-based, and nouvelle AI. Researchers from the related field of robotics, such as Rodney Brooks, rejected symbolic AI and focused on the basic engineering problems that would allow robots to move and survive. Their work revived the non-symbolic viewpoint of the early cybernetics researchers of the 1950s and reintroduced the use of control theory in AI. This coincided with the development of the embodied mind thesis in the related field of cognitive science: the idea that aspects of the body (such as movement, perception and visualization) are required for higher intelligence.

Computational intelligence and soft computing

Interest in neural networks and "connectionism" was revived by David Rumelhart and others in the middle of 1980s. Neural networks are an example of soft computing—they are solutions to problems which cannot be solved with complete logical certainty, and where an approximate solution is often sufficient. Other soft computing approaches to AI include fuzzy systems, evolutionary computation and many statistical tools. The application of soft computing to AI is studied collectively by the emerging discipline of computational intelligence.

Integrating the approaches

Intelligent agent paradigm

An intelligent agent is a system that perceives its environment and takes actions which maximize its chances of success. The simplest intelligent agents are programs that solve specific problems. More complicated agents include human beings and organizations of human beings (such as firms). The paradigm gives researchers license to study isolated problems and find solutions that are both verifiable and useful, without agreeing on one single approach. An agent that solves a specific problem can use any approach that works—some agents are symbolic and logical, some are sub-symbolic neural networks and others may use new approaches. The paradigm also gives researchers a common language to communicate with other fields—such as decision theory and economics—that also use concepts of abstract agents. The intelligent agent paradigm became widely accepted during the 1990s.

Agent architectures and cognitive architectures

Researchers have designed systems to build intelligent systems out of interacting intelligent agents in a multi-agent system. A system with both symbolic and sub-symbolic components is a hybrid intelligent system, and the study of such systems is artificial intelligence systems integration. A hierarchical control system provides a bridge between sub-symbolic AI at its lowest, reactive levels and traditional symbolic AI at its highest levels, where relaxed time constraints permit planning and world modelling. Rodney Brooks' subsumption architecture was an early proposal for such a hierarchical system.

.End.

参考译文 快速阅读1 人工智能研究的方法

目前还没有一个统一的理论或范式来指导人工智能研究。研究人员在许多问题上存在分歧。一些长期悬而未决的问题是：人工智能是否应该通过研究心理学或神经学来模拟自然智能？或者，人类生物学与人工智能研究无关，就像鸟类生物学与航空工程无关一样？智能行为可以用简单、优雅的原则（如逻辑或优化）来描述吗？还是一定要解决大量完全无关的问题？智力能用的类似于文字和思想高级符号再现吗，还是需要“次符号”处理？创造了GOFAI一词的约翰·豪格兰（John Haugeland）还提出，人工智能应该更恰当地称为合成智能，这一术语后来被一些非GOFAI的研究人员采用。

Stuart Shapiro将人工智能研究分为三种方法，他称之为计算心理学、计算哲学和计算机科学。计算心理学被用来制作模拟人类行为的计算机程序。计算哲学，是用来发展一个自适应的，自由流动的计算机思维。实现计算机科学的目标是创造出能够执行只有人类才能完成的任务的计算机。人类的行为、思维和动作共同构成了人工智能。

控制论与脑模拟

在20世纪40年代和50年代，一些研究人员探索了神经学、信息论和控制论之间的联系。他们中的一些人建造了利用电子网络来展示原始智能的机器，比如W.格雷·尔特的海龟和约翰·霍普金斯的野兽。这些研究人员中的许多人聚集在普林斯顿大学的目的论学会和英国的比率俱乐部的会议上。到1960年，这种方法基本上被放弃了，尽管其中的一些内容将在20世纪80年代又重新出现。

象征的

20世纪50年代中期，当数字计算机成为可能时，人工智能研究开始探索人类智能可以简化为符号操纵的可能性。这项研究集中在三个机构：卡内基梅隆大学、斯坦福大学和麻省理工学院，每个机构都有自己的研究风格。John Haugeland将这些人工智能方法命名为“老式A”

或“GOFAI”。在20世纪60年代，符号方法在小型演示程序中模拟高级思维方面取得了巨大成功。基于控制论或神经网络的方法被抛弃或被推到了后台。20世纪60年代和70年代的研究人员相信，符号方法最终会成功地创造出一台具有人工通用智能的机器，并认为这是他们领域的目标。

认知模拟

经济学家赫伯特·西蒙（Herbert Simon）和艾伦·纽威尔（Allen Newell）研究了人类解决问题的技能，并试图将其形式化，他们的工作奠定了人工智能领域以及认知科学、运筹学和管理科学的基础。他们的研究小组利用心理学实验的结果来开发程序，模拟人们用来解决问题的技术。这一传统以卡内基梅隆大学为中心，最终在20世纪80年代中期Soar建筑的发展中达到顶峰。

基于逻辑的

与Newell和Simon不同，John McCarthy认为机器不需要模拟人类的思维，而是应该尝试寻找抽象推理和解决问题的本质，而不管人们是否使用相同的算法。他在斯坦福大学（SAIL）的实验室专注于使用形式逻辑来解决各种各样的问题，包括知识表示、计划和学习。逻辑学也是爱丁堡大学和欧洲其他地方工作的重点，这导致了编程语言Prolog和逻辑编程科学的发展。

反逻辑还是邋遢

麻省理工学院的研究人员（如马文·明斯基和西摩·帕普特）发现，解决视觉和自然语言处理中的难题需要特殊的解决方案——他们认为，没有简单而通用的原则（如逻辑）能够捕捉到智能行为的所有方面。Roger Schank将他们的“反逻辑”方法描述为“邋遢”（与CMU和斯坦福的“整洁”范式相反）。常识知识库（如Doug Lenat的Cyc）是“邋遢”人工智能的一个例子，因为它们必须手工构建，一次只能创建一个复杂的概念。

基于知识

1970年左右，当拥有大量内存的计算机问世时，来自这三种传统的研究人员开始将知识融入人工智能应用程序。这场“知识革命”导致了专家系统的开发和部署（由爱德华·费根鲍姆介绍），这是第一个真正成功的人工智能形式软件。许多简单的人工智能应用程序需要大量的知识，这也推动了知识革命。

次符号

到了20世纪80年代，符号人工智能的发展似乎停滞不前，许多人认为符号系统永远无法模仿人类认知的所有过程，特别是感知、机器人、学习和模式识别。许多研究人员开始研究解决特定人工智能问题的“次符号化”方法。次符号方法设法在没有具体知识表示的情况下接近智能。

具体化的智能

这包括具体化的、情境化的、基于行为的和新的人工智能。来自机器人学相关领域的研究人员，如罗德尼·布鲁克斯，拒绝了象征性人工智能，而把重点放在让机器人移动和生存的基本工程问题上。他们的工作恢复了20世纪50年代早期控制论研究人员的非符号观点，并重新引入了控制理论在人工智能中的应用。这与认知科学相关领域中的体现心灵理论的发展相吻合：

身体的各个方面（如运动、感知和视觉化）都需要更高的智力。

计算智能与软计算

20世纪80年代中期，大卫·鲁梅尔哈特和其他人重新引起了对神经网络和“连接主义”的兴趣。神经网络是软计算的一个例子——它们是解决不能完全逻辑确定地解决的问题的方法，并且近似解往往是足够的。人工智能的其他软计算方法包括模糊系统、进化计算和许多统计工具。软计算在人工智能中的应用是由新兴的计算智能学科共同研究的。

整合方法

智能代理范式

智能代理是一个系统，它能感知环境并采取行动，最大限度地提高成功的机会。最简单的智能代理是解决特定问题的程序。更复杂的代理人包括人和人的组织（如公司）。这种范式允许研究人员研究孤立的问题，找到既可验证又有用的解决方案，而无须就单一方法达成一致。一个解决特定问题的代理可以使用任何有效的方法——一些代理是符号化和逻辑化的，一些是亚符号神经网络，还有一些可能使用新的方法。该范式还为研究人员提供了一种与其他领域（如决策理论和经济学）交流的通用语言，这些领域也使用抽象主体的概念。智能代理范式在20世纪90年代被广泛接受。

代理体系结构与认知体系结构

研究人员设计了一些系统，用多智能体系统中相互作用的智能体构建智能系统。具有符号和子符号成分的系统是一个混合智能系统，对这类系统的研究就是人工智能系统集成。分层控制系统在最低的次符号人工智能、反应性人工智能和最高水平的传统符号人工智能之间架起了一座桥梁，在那里放松的时间限制允许规划和世界建模。Rodney Brooks的包容体系结构是这种等级体系的早期建议。

Fast Reading Two | The Tools of Artificial Intelligence

Fast Reading Two

In the course of 50 years of research, AI has developed a large number of tools to solve the most difficult problems in computer science. A few of the most general of these methods are discussed below.

Search and optimization

Many problems in AI can be solved in theory by intelligently searching through many possible solutions: Reasoning can be reduced to performing a search. For example, logical proof can be viewed as searching for a path that leads from premises to conclusions, where each step is the application of an inference rule. Planning algorithms search through trees of goals and subgoals, attempting to find a path to a target goal, a process called means-ends analysis. Robotics algorithms for moving limbs and grasping objects use local searches in configuration space. Many learning algorithms use search algorithms based on optimization.

Logic

Logic is used for knowledge representation and problem solving, but it can be applied to other problems as well. For example, the satplan algorithm uses logic for planning and inductive logic programming is a method for learning.

Several different forms of logic are used in AI research. Propositional or sentential logic is the logic of statements which can be true or false. First-order logic also allows the use of quantifiers and predicates, and can express facts about objects, their properties, and their relations with each other. Fuzzy logic, is a version of first-order logic which allows the truth of a statement to be represented as a value between 0 and 1, rather than simply True (1) or False (0). Fuzzy systems can be used for uncertain reasoning and have been widely used in modern industrial and consumer product control systems. Subjective logic models uncertainty in a different and more explicit manner than fuzzy-logic: a given binomial opinion satisfies belief + disbelief + uncertainty = 1 within a Beta distribution. By this method, ignorance can be distinguished from probabilistic statements that an agent makes with high confidence.

Default logics, non-monotonic logics and circumscription are forms of logic designed to help with default reasoning and the qualification problem. Several extensions of logic have been designed to handle specific domains of knowledge, such as: description logics; situation calculus, event calculus and fluent calculus (for representing events and time); causal calculus; belief calculus; and modal logics.

Probabilistic methods for uncertain reasoning

Many problems in AI (in reasoning, planning, learning, perception and robotics) require the agent to operate with incomplete or uncertain information. AI researchers have devised a number of powerful tools to solve these problems using methods from probability theory and economics.

Bayesian networks are a very general tool that can be used for a large number of problems: reasoning (using the Bayesian inference algorithm), learning (using the expectation-maximization algorithm), planning (using decision networks) and perception (using dynamic Bayesian networks). Probabilistic algorithms can also be used for filtering, prediction, smoothing and finding explanations for streams of data, helping perception systems to analyze processes that occur over time (e.g., hidden Markov models or Kalman filters).

A key concept from the science of economics is “utility”: a measure of how valuable something is to an intelligent agent. Precise mathematical tools have been developed that analyze how an agent can make choices and plan, using decision theory, decision analysis, and information value theory. These tools include models such as Markov decision processes, dynamic decision networks, game theory and mechanism design.

Classifiers and statistical learning methods

The simplest AI applications can be divided into two types: classifiers ("if shiny then diamond") and controllers ("if shiny then pick up"). Controllers do, however, also classify conditions before inferring actions, and therefore classification forms a central part of many AI systems. Classifiers are functions that use pattern matching to determine a closest match. They can be tuned according to examples, making them very attractive for use in AI. These examples are known as observations or patterns. In supervised learning, each pattern belongs to a certain predefined class. A class can be seen as a decision that has to be made. All the observations combined with their class labels are known as a data set. When a new observation is received, that observation is classified based on previous experience.

A classifier can be trained in various ways; there are many statistical and machine learning approaches. The most widely used classifiers are the neural network, kernel methods such as the support vector machine, k-nearest neighbor algorithm, Gaussian mixture model, naive Bayes classifier, and decision tree.The performance of these classifiers have been compared over a wide range of tasks. Classifier performance depends greatly on the characteristics of the data to be classified. There is no single classifier that works best on all given problems; this is also referred to as the "no free lunch" theorem. Determining a suitable classifier for a given problem is still more an art than science.

Neural networks

A neural network is an interconnected group of nodes, akin to the vast network of neurons in the human brain.

The study of non-learning artificial neural networks began in the decade before the field of AI research was founded, in the work of Walter Pitts and Warren McCullouch. Frank Rosenblatt invented the perceptron, a learning network with a single layer, similar to the old concept of linear regression. Early pioneers also include Alexey Grigorevich Ivakhnenko, Teuvo Kohonen, Stephen Grossberg, Kunihiko Fukushima, Christoph von der Malsburg, David Willshaw, Shun-Ichi Amari, Bernard Widrow, John Hopfield, Eduardo R. Caianiello, and others.

The main categories of networks are acyclic or feedforward neural networks (where the signal passes in only one direction) and recurrent neural networks (which allow feedback and short-term memories of previous input events). Among the most popular feedforward networks are perceptrons, multi-layer perceptrons and radial basis networks. Neural networks can be applied to the problem of intelligent control (for robotics) or learning, using such techniques as Hebbian learning, GMDH or competitive learning.

Control theory

Control theory, the grandchild of cybernetics, has many important applications, especially in robotics.

Languages

AI researchers have developed several specialized languages for AI research, including Lisp and Prolog.

.End.

参考译文 快速阅读2 人工智能研究的工具

在50年的研究过程中，人工智能开发了大量的工具来解决计算机科学中最困难的问题。下面将讨论其中最普遍的几种。

搜索和优化

人工智能中的许多问题在理论上可以通过智能搜索来解决。例如，逻辑证明可以看作是寻找一条从前提到结论的路径，其中每一步都是推理规则的应用。规划算法在目标树和子目标树中搜索，试图找到一条通向目标的路径，这个过程称为手段-终点分析。机器人学中用于移动肢体和抓取物体的算法使用配置空间中的局部搜索。许多学习算法使用基于优化的搜索算法。

逻辑

逻辑用于知识表示和问题解决，但也可以应用于其他问题。例如，satplan算法使用逻辑进行规划，归纳逻辑编程是一种学习方法。

人工智能研究中使用了几种不同形式的逻辑。命题逻辑或句子逻辑是陈述的逻辑，可以是真的也可以是假的。一阶逻辑还允许使用量词和谓词，并且可以表达关于对象、它们的属性以及它们之间的关系的事实。模糊逻辑是一阶逻辑的一个版本，它允许语句的真值表示为0到1之间的值，而不是简单的真（1）或假（0）。模糊系统可以用于不确定推理，在现代工业和消费品控制系统中得到了广泛的应用。主观逻辑以不同于模糊逻辑的方式对不确定性进行建模：给定的二项式观点在β分布中满足信念+不信任+不确定性=1。通过这种方法，可以将无知与智能体高置信度的概率陈述区分开来。

缺省逻辑、非单调逻辑和限定是逻辑的形式，旨在帮助解决缺省推理和限定问题。逻辑的几个扩展被设计来处理特定的知识领域，例如：描述逻辑；情景演算、事件演算和流畅演算（用于表示事件和时间）；因果演算；信念演算；模态逻辑。

不确定推理的概率方法

人工智能中的许多问题（推理、规划、学习、感知和机器人技术）都要求智能体在不完全或不确定的信息下进行操作。人工智能的研究人员已经设计了许多强大的工具来解决这些问题，使用的方法来自概率论和经济学。

贝叶斯网络是一个非常通用的工具，可以用于许多问题：推理（使用贝叶斯推理算法）、学习（使用期望最大化算法）、规划（使用决策网络）和感知（使用动态贝叶斯网络）。概率算法也可用于过滤、预测、平滑和寻找数据流的解释，帮助感知系统分析随时间发生的过程（例如，隐马尔可夫模型或卡尔曼滤波器）。

经济学中的一个关键概念是“效用”：衡量某物对智能体有多大价值。利用决策理论、决策分析和信息价值理论，已经开发出了精确的数学工具来分析智能主体如何做出选择和计划。这些工具包括马尔可夫决策过程、动态决策网络、博弈论和机制设计等模型。

分类器与统计学习方法

最简单的人工智能应用程序可以分为两种类型：分类器（“如果有光泽，则为菱形”）和控制器（“如果闪亮则提取”）。然而，控制器在推断动作之前也会对情况进行分类，因此分类是许多人工智能系统的核心部分。分类器是使用模式匹配来确定最接近匹配的函数。它们可以根据示例进行调整，这使得它们在人工智能中的使用非常有吸引力。这些例子被称为观察或模式。在监督学习中，每一个模式都属于一个预先定义的类。一个类可以看作是一个必须做出的决定。所有的观察值和它们的类标签一起被称为一个数据集。当收到新的观察结果时，该观察结果将根据以前的经验进行分类。

分类器可以通过多种方式进行训练；有许多统计和机器学习方法。目前应用最广泛的分类器有神经网络、支持向量机、k近邻算法、高斯混合模型、朴素贝叶斯分类器和决策树。这些分类器的性能已经在一系列任务中进行了比较。分类器的性能在很大程度上取决于待分类数据的特征。没有一个分类器对所有给定的问题都能起到最好的效果；这也被称为“没有免费午餐”定理。为给定的问题确定合适的分类器仍然是一门艺术而不是科学。

神经网络

神经网络是一组相互连接的节点，类似于人脑中庞大的神经元网络。

非学习人工神经网络的研究始于人工智能研究领域建立之前的十年，由沃尔特·皮茨和沃伦·麦卡鲁奇完成。弗兰克·罗森布拉特发明了感知器，一种单层的学习网络，类似于线性回归的旧概念。早期先驱者还包括亚历克赛·格里戈列维奇·伊瓦克年科、特沃·科霍宁、斯蒂芬·格罗斯伯格、福岛久尼希科、克里斯托夫·冯·德马尔斯堡、大卫·威尔肖、顺一阿玛里、伯纳德·维德罗、约翰·霍普菲尔德、爱德华多·R·卡亚尼埃洛等。

网络的主要类别有无环或前馈神经网络（其中信号只沿一个方向传递）和递归神经网络（允许对先前输入事件进行反馈和短期记忆）。最流行的前馈网络有感知器、多层感知器和径向基网络。利用Hebbian学习、GMDH或竞争学习等技术，神经网络可以应用于智能控制或学习问题。

控制论

控制理论作为控制论的下位概念，有着许多重要的应用，特别是在机器人学中。

语言

人工智能研究人员已经为人工智能研究开发了几种专门的语言，包括Lisp和Prolog。

Ex 1 **What is the AI? Try to give a brief summary according to the passage.**

Ex 2 **Fill in the table below by matching the corresponding Chinese or English equivalents.**

AI	
	人工的
stocks	
	智力
mimic	
	仿真

Ex 3 **Choose the best answer to the following statements according to the text we've learnt.**

1. ________ (AI), sometimes called machine intelligence, is intelligence demonstrated by machines, unlike the natural intelligence displayed by humans and animals.
   A. Interfaces B. Programs
   C. Kernel techniques D. Artificial intelligence
2. Artificial intelligence sometimes called ________ is intelligence demonstrated by machines, unlike the natural intelligence displayed by humans and animals.
   A. machine intelligence B. intelligence
   C. Graphical User Interface D.AI
3. Leading AI textbooks define the field as the study of " ________": any device that perceives its environment and takes actions that maximize its chance of successfully achieving its goals.
   A. AI B. intelligent agents
   C. Windows D. machine intelligence
4. Colloquially, the term "artificial intelligence" is often used to describe ________ (or computers) that mimic "cognitive" functions that humans associate with the human mind, such as "learning" and "problem solving".
   A. platform B. Plug and Play
   C. intelligent agents D. machines
5. As machines become increasingly capable, tasks considered to require "intelligence" are often removed from the definition of AI, a phenomenon known as the ________.
   A. AI effect B. intelligent agents
   C. machine intelligence D. machines

Part B Practical Learning

Task One

Task Two

Training Target

In this part, there are two tasks in English environment. You should complete these tasks in groups under the joint guidance of professional teachers and laboratory teachers, so as to train and improve your ability to complete professional tasks in English environment.

In this part, the students must finish two special tasks in English environment, Under the guidance of the specialized English teacher and the teacher teaching the network query in the computer laboratory. The students must work with their students in the same group.

Task One Search for Cases of Artificial Intelligence on the Internet

In this task, students look for cases about artificial intelligence on the Internet, and understand the application of artificial intelligence in life.

Here are some examples of finding artificial intelligence related information. I hope this information can help students complete the task. (Pic 10.1, Pic10.2)

Pic 10.1 Example of finding artificial intelligence related information (1)

Pic 10.2 Example of finding artificial intelligence related information (2)

Task Two Complete an Application Related to Artificial Intelligence

In this task, students need to complete an application or operation related to artificial intelligence. For example, remote light on, intelligent rice cooker to make an appointment for cooking, or use intelligent washing machine to wash clothes at fixed points, and sweeping robot can also be used to do cleaning work, and so on. You can feel the convenience and high mechanization of artificial intelligence. (Pic 10.3)

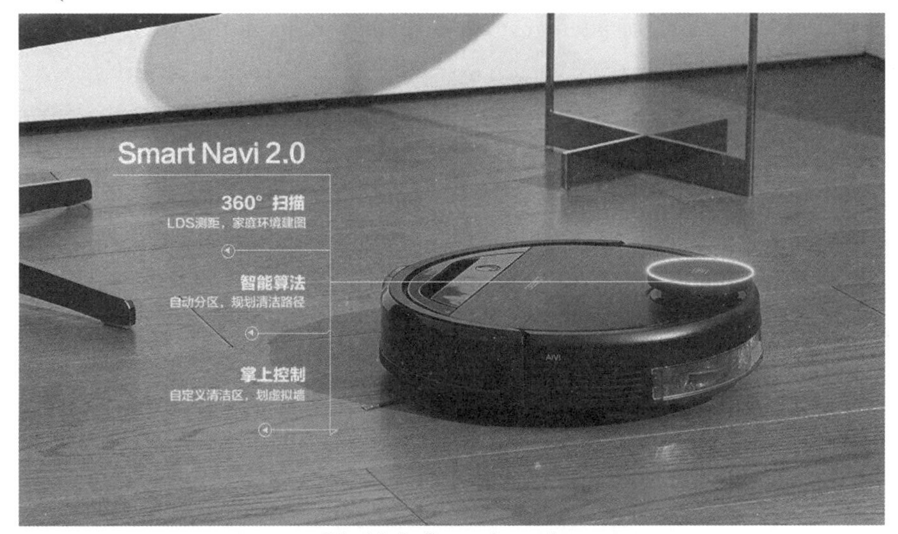

Pic 10.3 Sweeping robot

Part C Occupation English

Occupation English

In this part, there is an English dialogue in real life and work environment. You will play the roles of A and B and read the dialogue aloud to practice your ability to use English.

Intelligent Car

智能汽车

Role setting: employees of an automobile enterprise (A) and employees of an automobile enterprise (B)

角色设置：某汽车企业员工（A），某汽车企业员工（B）

A: Have you heard that Baidu's driverless cars have been tested in Guangzhou, Beijing and other places?你听说百度的无人驾驶汽车已经在广州，北京等地测试上路了吗？

B: Of course, it's hard to know whether you want to know. Nowadays, the means of information transmission is more and more fast. There are news about this on the Internet and photos of people trying to ride.当然了，想不知道都难，现在信息传递的手段多还快，网上到处是这方面的消息，还有市民试乘的照片呢。

A: No, it's not. The technology is really advanced now. It doesn't need to be controlled by people at all. The car can travel freely on the road. This is something that I didn't dare to think of before.可不是，现在的技术真先进啊，完全不用人控制，车自己可以在路上自由穿行，这是以往都不敢想的事。

B: Driverless vehicle is a kind of intelligent car, also known as wheeled mobile robot. It mainly depends on the intelligent driver of computer system in the car to achieve driverless.无人驾驶汽车是智能汽车的一种，也称为轮式移动机器人，主要依靠车内的以计算机系统为主的智能驾驶仪来实现无人驾驶的。

A: I know that it uses vehicle sensors to sense the surrounding environment of the vehicle, and controls the steering and speed of the vehicle according to the road, vehicle position and obstacle information obtained by the perception, so that the vehicle can drive safely and reliably on the road.我知道它是利用车载传感器来感知车辆周围环境，并根据感知所获得的道路、车辆位置和障碍物信息，控制车辆的转向和速度，从而使车辆能够安全、可靠地在道路上行驶。

B: Think about the smart cars running all over the street in the future. What a scene it is! It's exciting to think about it.想想以后满大街跑的都是智能汽车，那是怎样的一副场景！想想就让人兴奋。

A: That's right. The development of technology and social progress can't be stopped. A better life beckons to us.是的，技术的发展，社会的进步是不可阻止的，美好的生活再向我们招手。

Word Building

前缀/后缀由一个或几个字母组成，放在词根或单词之前/之后，组成一个新词。

(1) -ance, -ence（后缀）：表示“性质，状况，行为，过程，总量，程度”

import 进口 ———— importance 重要

diligent 勤勉的 ———— diligence 勤奋

(2) -ic（后缀）：表示“……学……法”

log 说话 ———— logic 逻辑

electron 电子 ———— electronics 电子学

(3) -fy（后缀）：表示“使……化,使成”

pure 纯洁 ———— purify 净化

beauty 美 ———— beautify 美化

(4) -ly（后缀）：形容词变副词 ly

bad 坏的 ———— badly 拙劣地

careful 仔细的 ———— carefully 仔细地

(5) dis-（前缀）：表示否定“不”

like 喜欢 ———— dislike 不喜欢

order 秩序 ———— disorder 混乱

Ex Translate the following words and try your best to guess the meaning of each word on the right according to the clues given on the left.

different	不同(形容词)	difference________
obey	服从(动词)	obedience________
linguist	语言学家(名词)	linguistics________
machine	机器(名词)	mechanics________
intense	强烈的(形容词)	intensify________
sign	信号(名词)	signify________
beautiful	美丽的(形容词)	beautifully ________
healthy	健康(名词)	healthily________
close	关闭（动词)	disclose________
appear	出现(动词)	disappear________

Exercises

Ex 1 What is AI ? Try to give a brief summary of this passage in no more than five sentences.

Ex 2 How many aspects of the application of artificial intelligence are there? Try to give several examples.

Ex 3 Fill in the table below by matching the corresponding Chinese or English equivalents.

AI	
	延迟
data-mining	
	仿真
automotive	
	刹车
healthcare	
	感知
vaccine	
	垃圾邮件

Ex 4 Choose the best answer for each of the following statements according to the text we've learnt.

1. ______________ is relevant to any intellectual task.

   A. Tools

   B. Rules

   C. AI

   D. Instructions

2. Artificial intelligence is breaking into the healthcare industry by______________.

   A. machine

   B. assisting doctors

   C. doctor

   D. people

3. Many components contribute to the functioning of ______ cars.

   A. driver

   B. self-driving

   C. map

   D. application

4. One main factor that influences the ability for a driver-less car to function is ____________.

A. mapping

B. navigation

C. Smart Tags

D. brake

5. Banks use ______________ systems to organize operations, maintain book-keeping, invest in stocks, and manage properties.

A. artificial intelligence

B. intelligence

Ç. artificial

D. driver

Reference

[1] Timothy J.O'Leary ,Linda I.O'Leary.计算机专业英语- Computing Essentials.[M].北京：高等教育出版社，2008.

[2] 金志权. 计算机专业英语教程[M].7版.北京：电子工业出版社，2020.

[3] 卜艳萍，周伟. 计算机专业英语教程[M].3版.北京：清华大学出版社，2019.

[4] 柯晓华.计算机专业英语教程[M].北京：科学出版社，2017.